BIOTECHNOLOGY
INTELLIGENCE
UNIT

CRYSTALLINE BACTERIAL CELL SURFACE PROTEINS

Uwe B. Sleytr

Paul Messner

Dietmar Pum

Margit Sára

Center for Ultrastructure Research and
Ludwig Boltzmann Institute for Molecular Nanotechnology
Universität für Bodenkultur
Vienna, Austria

Academic Press

R.G. LANDES COMPANY
AUSTIN

Biotechnology Intelligence Unit
Crystalline Bacterial Cell Layer Proteins

R.G. LANDES COMPANY
Austin, Texas, U.S.A.

Submitted: October 1995
Published: January 1996

Please address all inquiries to the Publisher:
R.G. Landes Company
909 Pine Street, Georgetown, Texas, U.S.A. 78626
Phone: 512/ 863 7762; FAX: 512/ 863 0081

Academic Press, Inc.
525 B Street, Suite 1900, San Diego, California, U.S.A. 92101-4495

United Kingdom Edition published by Academic Press Limited
24-28 Oval Road, London NW1 7DX, United Kingdom

International Standard Book Number (ISBN): 0-12-648470-8
Library of Congress Catalog Number: not available by publication date

Library of Congress Cataloging-in-Publication Data

(CIP data applied for but not available by publication date)

Transferred to Digital Printing, 2011

Printed and bound in the United Kingdom

PUBLISHER'S NOTE

R.G. Landes Company publishes six book series: *Medical Intelligence Unit, Molecular Biology Intelligence Unit, Neuroscience Intelligence Unit, Tissue Engineering Intelligence Unit, Environmental Intelligence Unit* and *Biotechnology Intelligence Unit.* The authors of our books are acknowledged leaders in their fields and the topics are unique. Almost without exception, no other similar books exist on these topics.

Our goal is to publish books in important and rapidly changing areas of bioscience for sophisticated researchers and clinicians. To achieve this goal, we have accelerated our publishing program to conform to the fast pace in which information grows in bioscience. Most of our books are published within 90 to 120 days of receipt of the manuscript. We would like to thank our readers for their continuing interest and welcome any comments or suggestions they may have for future books.

Deborah Muir Molsberry
Publications Director
R.G. Landes Company

CONTENTS

EDITORS

Uwe B. Sleytr
Center for Ultrastructure Research
and Ludwig Boltzmann Institute for Molecular Nanotechnology
Universität für Bodenkultur
Vienna, Austria
Chapters 1, 2, 6, 7, 8

Paul Messner
Center for Ultrastructure Research
and Ludwig Boltzmann Institute for Molecular Nanotechnology
Universität für Bodenkultur
Vienna, Austria
Chapters 1, 2, 3, 7

Dietmar Pum
Center for Ultrastructure Research
and Ludwig Boltzmann Institute for Molecular Nanotechnology
Universität für Bodenkultur
Vienna, Austria
Chapters 1, 2, 8

Margit Sára
Center for Ultrastructure Research
and Ludwig Boltzmann Institute for Molecular Nanotechnology
Universität für Bodenkultur
Vienna, Austria
Chapters 1, 2, 5, 6

CONTRIBUTORS

Eva-Maria Egelseer
Center for Ultrastructure Research
and Ludwig Boltzmann Institute for Molecular Nanotechnology
Universität für Bodenkultur
Vienna, Austria
Chapter 5

Beatrix Kuen
Institute for Microbiology and Genetics
Universität Wien
Vienna, Austria
Chapter 4

Seta Küpcü
Center for Ultrastructure Research and
Ludwig Boltzmann Institute for Molecular Nanotechnology
Universität für Bodenkultur
Vienna, Austria
Chapter 6

Werner Lubitz
Institute for Microbiology and Genetics
Universität Wien
Vienna, Austria
Chapter 4

Frank M. Unger
Center for Ultrastructure Research and
Ludwig Boltzmann Institute for Molecular Nanotechnology
Universität für Bodenkultur
Vienna, Austria
Chapter 7

PREFACE

Most prokaryotic cells possess layered assemblies of homo- and heteropolymers external to the cytoplasmic membrane which function as important interface between the environment and the cell. As such, the supramolecular architecture of envelopes represents very specific evolutionary adaptations of unicellular life forms to different environmental conditions and selection criteria. Although bacterial cell envelope structures are one of the most intensively studied major structures of microbial cells it took relatively long until it became evident that monomolecular arrays of protein or glycoprotein subunits (S-layers) are one of the most common surface structures found in prokaryotic organisms. The aim of this book is to assemble our present day understanding of the occurrence, structure, chemistry, genetics, assembly, function and application potential of S-layers. Each chapter is designed in a way that it stands as a self-contained unit. We hope that this book will help to stimulate further development in basic and applied S-layer research.

Uwe B. Sleytr
Paul Messner
Dietmar Pum
Margit Sára
October 1995

ACKNOWLEDGMENT

The editors wish to thank the contributors to this book for delivering the manuscripts in time, which is considered an exception to the rule. Although many more individuals helped in completing this book Helene Hendling, Dieter Jäger and Andrea Scheberl deserve particular praise for all their help in finishing the project.

CHAPTER 1

INTRODUCTION

Uwe B. Sleytr, Paul Messner, Dietmar Pum, Margit Sára

Crystalline bacterial cell surface layers composed of proteinaceous subunits, commonly referred to as S-layers,[1,2] were relatively unknown 25 years ago. Their identification on the cell surface of *Acinetobacter, Aquaspirillum* (*Spirillum*), *Bacillus* and *Clostridium* species was considered to represent a rather unique cell wall structure.[3-8] In a review in 1978[1] already 80 species could be listed and by 1992 more than 300 organisms representing some 93 genera of prokaryotes were known to possess crystalline surface layers.[9]

S-layers now can be considered as one of the most commonly observed bacterial cell surface structures (see appendix, page 211). With a few exceptions the cell wall of archaeobacterial cells consists exclusively of crystalline surface layers. S-layers were already detected in hundreds of different species of nearly every taxonomical group of walled eubacteria. Since it was shown that S-layers of eubacteria tend to be lost during cultivation of the organisms under laboratory conditions, it was only a few years ago that their wide-spread occurrence in fresh isolates was appreciated.

Morphological, chemical, genetic and morphogenetic studies clearly demonstrated that S-layers represent the simplest type of biological membrane developed during evolution of prokaryotic cells. Most S-layers are composed of a single protein or glycoprotein species endowed with the ability to assemble into a monomolecular lattice by an entropy-driven process. S-layers are

Crystalline Bacterial Cell Surface Proteins, edited by Uwe B. Sleytr, Paul Messner, Dietmar Pum, Margit Sára. © 1996 R.G. Landes Company.

also highly porous structures which completely cover the cell surface during all stages of cell growth and cell division.

Since S-layer-carrying organisms were demonstrated to be ubiq-
uitous in the biosphere and because S-layers represent one of the
most abundant of cellular proteins, it became obvious that these
metabolically expensive products must provide the organism with
an advantage of selection in very different habitats. Although, many
of the functions assigned to S-layers are still hypothetical, it is
now recognized that they can provide the organism with a selection advantage by functioning as protective coats, molecular sieves,
molecule and ion traps and as structures involved in cell adhesion
and surface recognition. S-layers were also identified as contribut-
ing to virulence when present on pathogenic organisms. In those
archaeobacteria which possess crystalline arrays as exclusive cell wall
component, S-layers act as a framework that determines cell shape
and as a structure aiding in the cell division process.

Since S-layers represent a very important class of secreted
(glyco)proteins and possess a high degree of structural regularity,
they represent interesting model systems for studies on structure,
biosynthesis, genetics, glycosylation, functions and dynamic aspects
of assembly of exoproteins as well as evolutionary relationship and
specific adaptations within the prokaryotic world.[1,2,9-19]

The wealth of information existing on the general principles
of crystalline bacterial cell surface layers more recently has revealed
a broad application potential in biotechnology, vaccine development, diagnostics, molecular nanotechnology and biomimetics. This
volume shall provide a detailed survey of the whole field of basic
and applied S-layer research. It is particularly hoped, that this presentation encourages new scientists to enter this interesting
interdisciplinary subject.

REFERENCES
    1. Sleytr UB. Regular arrays of macromolecules on bacterial cell walls:
       structure, chemistry, assembly and function. Int Rev Cytol 1978;
       53:1-64.
    2. Sleytr UB, Messner P, Pum D, Sára M, eds. Crystalline Bacterial
       Cell Surface Layer. Berlin: Springer, 1988.
    3. Glauert AM, Thornley MJ. The topography of the bacterial cell
       wall. Annu Rev Microbiol 1969; 23:159-98.

4. Glauert AM, Thornley MJ, Thorne KJI et al. The surface structure of bacteria. In: Fuller RF, Lovelock DW, eds. Microbial Ultrastructure. London: Academic Press, 1976:31-47.

5. Thornley MJ, Glauert AM, Sleytr UB. Structure and assembly of bacterial surface layers composed of regular arrays of subunits. Phil Trans R Soc London B 1974; 268:147-53.

6. Sleytr UB, Adam H, Klaushofer H. Die Feinstruktur der Zellwand von zwei thermophilen Clostridienarten, dargestellt mit Hilfe der Gefrierätztechnik. Mikroskopie 1968; 23:1-10.

7. Murray RGE. On the cell wall structure of *Spirillum serpens*. Can J Microbiol 1963; 9:381-92.

8. Holt SC, Leadbetter ER. Comparative ultrastructure of selected aerobic spore-forming bacteria: a freeze etching study. Bact Rev 1969; 33:346-78.

9. Messner P, Sleytr UB. Crystalline bacterial cell surface layers. In: Rose AH, ed. Advances in Microbial Physiology. Vol 33. London: Academic Press, 1992:213-75.

10. Sleytr UB, Glauert AM. Bacterial cell walls and membranes. In: Harris JR, ed. Electron Microscopy of Proteins. Vol 3. London: Academic Press, 1982:41-76.

11. Sleytr UB, Messner P. Crystalline surface layers on bacteria. Annu Rev Microbiol 1983; 37:311-39.

12. Smit J. Protein surface layers of bacteria. In: Inouye M, ed. Bacterial Outer Membranes as Model Systems. New York: Wiley, 1987:343-76.

13. Koval SF. Paracrystalline protein surface arrays on bacteria. Can J Microbiol 1988; 34:407-414.

14. Sleytr UB, Messner P. Crystalline surface layers in prokaryotes. J Bacteriol 1988; 170:2891-97.

15. Hovmöller S, Sjögren A, Wang DN. The structure of crystalline bacterial surface layers. Prog Biophys Mol Biol 1988; 51:131-63.

16. Baumeister W, Wildhaber I, Phipps BM. Principles of organization in eubacterial and archaebacterial surface proteins. Can J Microbiol 1989; 35:215-27.

17. Beveridge TJ, Koval SF, eds. Advances in Bacterial Paracrystalline Surface Layers. New York: Plenum, 1993.

18. Sleytr UB, Messner P, Pum D et al. Crystalline bacterial cell surface layers. Mol Microbiol 1993; 10:911-16.

19. Beveridge TJ. Bacterial S-layers. Curr Opinion Struct Biol 1994; 4:204-12.

CHAPTER 2

OCCURRENCE, LOCATION, ULTRASTRUCTURE AND MORPHOGENESIS OF S-LAYERS

Uwe B. Sleytr, Paul Messner, Dietmar Pum, Margit Sára

2.1. INTRODUCTION

Chemical, ultrastructural and taxonomical studies revealed that in the course of evolution prokaryotic organisms have developed a broad spectrum of cell envelope structures. Despite this diversity, in general two separate surface enveloping structures can be distinguished, the plasma membrane and the associated cell wall proper.[1-4] Since bacteria are unicellular life forms, the different supramolecular architectures of the envelope function as important interface between the environment and the cell. As such, cell envelope structures also reflect evolutionary adaptations of the organism to specific environmental conditions and selection criteria. The envelope of a single-cell organism must serve multiple purposes. With a few exceptions it must be stable enough to withstand the turgor pressure of the protoplast and is directly involved in maintenance and determination of cell shape. The cell envelope has to allow nutrients to pass selectively into and waste products out of the cell but at the same time a relatively constant chemical composition inside the cell has to be maintained.

Crystalline Bacterial Cell Surface Proteins, edited by Uwe B. Sleytr,
Paul Messner, Dietmar Pum, Margit Sára. © 1996 R.G. Landes Company.

Our present knowledge of the molecular architecture of cell envelopes of prokaryotic organisms is based on results obtained by high resolution electron microscopy and biochemical and biophysical analysis primarily accomplished by subcellular fractionation.[5] Despite the fact, that considerable variations exist in the complexity and structure of bacterial cell envelopes, most envelope profiles can be classified into three main groups which also strongly reflect the phylogenetic position of the organisms and their Gram-staining reaction (Fig. 2.1). One of the most commonly observed cell surface structures on prokaryotes are two-dimensional arrays of proteinaceous subunits termed S-layers.[6,7] Crystalline proteinaceous layers similar to S-layer have been detected on the surface of the cell wall of eukaryotic algae[8] and in spore coats of endospores in *Bacillaceae*.[9]

2.2. OCCURRENCE AND LOCATION OF S-LAYERS ON BACTERIAL CELL ENVELOPES

S-layers represent an almost universal feature of arachaeobacteria[10-12] which represent a third line (domain) of evolutionary descent distinct from the eubacteria and eukaryotic cells.[13,14] Crystalline surface layers also have been observed in hundreds of different species of nearly every taxonomical group of walled eubacteria (an up-to-date list of S-layer carrying organisms is given in the appendix). Amongst archaeobacteria belonging to *Halobacteriales* (extreme halophiles), *Sulfolobales* and *Thermoproteales* (sulfur-dependent extreme thermophiles), *Thermococcales*, *Methanococcales* and *Methanomicrobiales* (methanogens), cell envelopes consist exclusively of an S-layer associated with the plasma membrane (Fig. 2.1a). In Gram-positive archaeobacteria (Fig. 2.1c) the cell envelope consists of a rigid cell wall sacculus which is apposed to the cytoplasmic membrane and an S-layer as outermost component. In *Methanothermus* and *Methanopyrus* this sacculus is composed of pseudomurein and in *Methanosarcina* of methanochondroitin. Even more complex envelope structures have been described for *Methanospirillum* and *Methanothrix*. These cells possess S-layers as (exclusive) wall components but are surrounded by a matrix layer of carbohydrate and protein components. The chain of cells is encased by a regularly structured protein sheath

(Fig. 2.1b). Individual cells are separated by spacer plugs composed of two different arrays of proteins. In some archaeobacteria two superimposed S-layers composed of different protein subunits have been observed. In Gram-positive eubacteria (Fig. 2.1c) the rigid sacculus is composed of peptidoglycan of variable thickness.

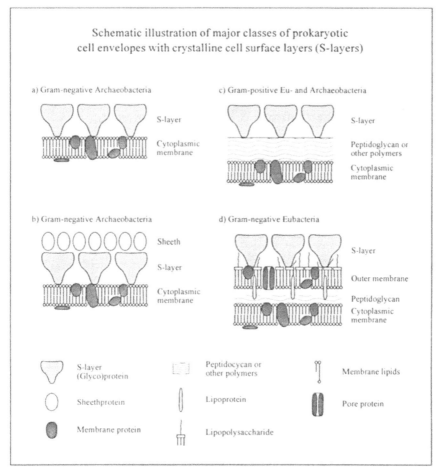

Fig. 2.1. Schematic illustration of the molecular architecture of the major classes of prokaryotic cell envelopes containing S-layers. (a) Cell envelope structure of Gram-negative archaeobacteria with crystalline S-layers as the only cell wall component external to the cytoplasmic membrane. (b) Gram-negative archaeobacterial cell envelope with an additional sheath composed of regularly arranged subunits. (c) The cell envelope as observed in Gram-positive eubacteria and archaeobacteria. In eubacteria the rigid wall component is primarily composed of peptidoglycan. In archaeobacteria other wall polymers (e.g. pseudomurein or methanochondroitin) are found. (d) In Gram-negative eubacteria the S-layer is closely associated with the outer membrane. (Modified after Sleytr UB, Messner P, Pum D et al. J Appl Bacteriol 1993; 74:21S-32S.)

Secondary cell wall polymers can be associated with this layer. Particularly with S-layer-carrying organisms belonging to the *Bacillaceae*, a relatively thin peptidoglycan layer could be observed[6,15] which may lead to Gram-variable or even Gram-negative staining reactions.[16] The most complex envelope structure can be observed in Gram-negative eubacteria (Fig. 2.1d). They consist of a thin peptidoglycan layer external to the cytoplasmic membrane and an outer membrane which is composed of lipopolysaccharides, phospholipids, proteins (e.g. porins) and lipoproteins. The latter function as a linker between the outer membrane and the peptidoglycan layer. If S-layers are present they are associated with components of the outer membrane. As observed with other types of envelopes, more than one S-layer may be present.

As described in detail in chapter 3, S-layers are generally composed of a single homogeneous protein or glycoprotein species with molecular weights ranging from 40,000 to 200,000. While the majority of archaeobacterial S-layer proteins appears to be glycosylated in eubacteria, this modification of the subunits up to now was only observed in *Bacillaceae*.[15,17-19] (See also chapter 3.)

2.3. ULTRASTRUCTURE OF S-LAYERS

The most useful electron microscopical preparation techniques for identifying S-layers on a particular organism are freeze-etching of intact cells (Fig. 2.2a-c), or negative staining of cell wall or envelope preparations.[20] Analyses of freeze-etched preparations of a great variety of archaeobacteria and eubacteria have shown that the crystalline arrays completely cover the cell surface at all stages of cell growth and division.[6,7,15,21-25] High resolution studies on the mass distribution of S-layers were generally performed on negatively stained preparations of isolated or recrystallized S-layers.[26-30] Both two- and three-dimensional computer image reconstruction techniques were applied for the evaluation of electron micrographs. With these computer image enhancement procedures, structural information on S-layer lattices down to approximately 1.5 nm have been revealed. S-layer lattices show oblique (p1, p2), square (p4) or hexagonal (p3, p6) symmetry (Figs. 2.3 and 2.4). Hexagonal symmetry of S-layers is predominant among the archaeobacteria (see appendix and ref. 17). Depending on the lattice type, the

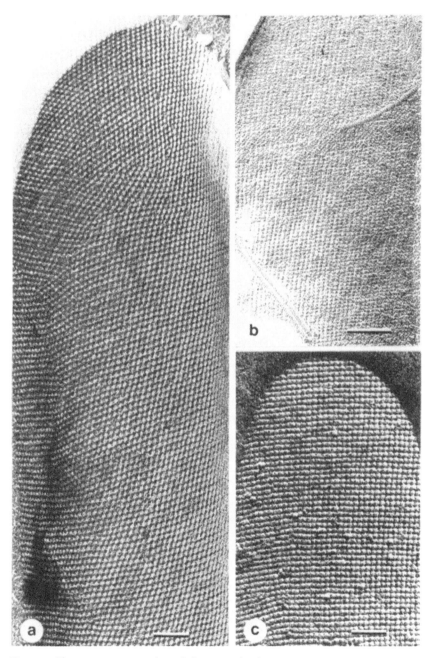

Fig. 2.2. Electron micrographs of freeze-etched preparations of intact cells: (a) Thermoanaerobacter thermohydrosulfuricus *L111-69 showing a hexagonal (p6) S-layer lattice which completely covers the cell surface. (b)* Bacillus stearothermophilus *NRS 2004/3a with an oblique (p2) S-layer. (c)* Desulfotomaculum nigrificans *NCIB 8706 revealing a square (p4) S-layer lattice. Bars = 100 nm.*

morphological units constituting the crystal lattice consist of one, two, three, four and six identical protein or glycoprotein subunits. However, some S-layer lattices apparently are composed of two subunit proteins with different molecular weights.[31-35] The morphological units may have center-to-center spacings of approximately 3 to 30 nm (see appendix). The protein or glycoprotein lattices are generally 5 to 25 nm thick. Amongst archaeobacterial S-layers frequently pillar-like domains on the inner surface can be observed which are associated with the plasma

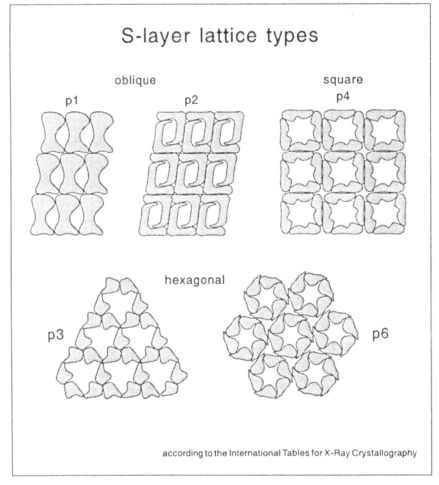

Fig. 2.3. Schematic illustration of the space groups found for S-layer lattices. The unit cells which are the building blocks of the lattices are composed of mono-, di-, tri-, tetra- or hexamers.

membrane. This characteristic structural feature, in some instances, may lead to S-layer-like structures which extend up to approximately 70 nm from the surface of the cytoplasmic membrane.[27,36]

High resolution studies revealed that pores of identical size and morphology exist between the regularly arranged S-layer subunits. In many S-layers, two or even more distinct classes of pores could be observed (Fig. 2.4). Pore sizes in the range from 2 nm to 8 nm and a porosity of the protein meshwork between 30% and 70%

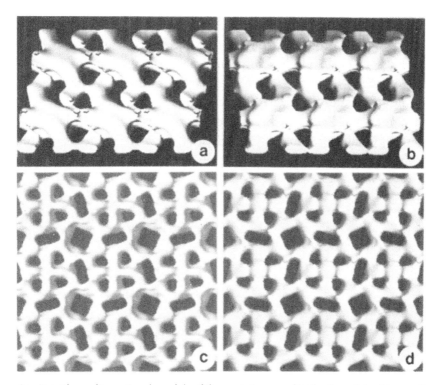

Fig. 2.4. Three-dimensional models of the protein mass distribution of the S-layer of Bacillus coagulans E38-66 (a) and of Bacillus stearothermophilus NRS 2004/3a (b). In both panels the left view shows the outer face and the right view the inner face. The morphological unit of Bacillus coagulans E38-66 shows oblique lattice symmetry (p2) with base vector lengths of 9.4 nm and 7.4 nm. The isoporous protein meshwork reveals an elongated and a round pore type. The S-layer is appr. 5 nm thick. Bacillus stearothermophilus NRS 2004/3a shows a square (p4) S-layer lattice with a center-to-center spacing between the morphological units of 13.5 nm, and a thickness of appr. 8 nm. The protein meshwork shows one large, two elongated and four small pores per morphological unit. The models were obtained after recording several tilt series in a transmission electron microscope, performing Fourier-domain computer image reconstructions over each individual tilted view, and combining all processed views to a three-dimensional data volume.

have been estimated from electron micrographs. Two- and three-dimensional images of S-layers obtained by computer image processing of electron micrographs reveal great variations in mass distribution, degree of handedness, and subunit-to-subunit linkages amongst different S-layers. A characteristic structural feature of many S-layer lattices is a rather smooth outer surface and a more corrugated inner surface.[26-28,30] More recently, S-layer lattices also have been studied by scanning probe microscopy.[28,37-39] Both scanning tunneling microscopy and atomic force microscopy have been applied. The topographical images obtained by both techniques strongly resemble the three-dimensional reconstructions of S-layers derived from electron microscopical images (Goessl A, Hödl C, Gross H et al. High resolution images of the surface topography of the S-layer of *Bacillus coagulans* E38-66/v1 obtained by atomic force and cryoelectron microscopy. Submitted). Particularly the underwater atomic force microscopy on unfixed S-layer lattices recrystallized onto a solid support (Fig. 2.5) revealed structural resolutions down to 1.2 nm. (See also chapter 8.)

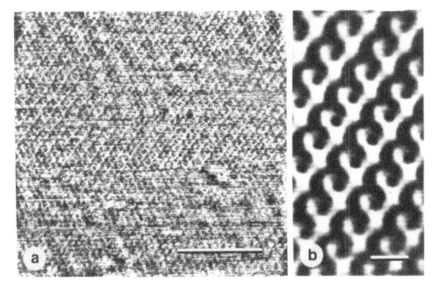

Fig. 2.5. (a) Scanning force microscopical image of the topography of the inner surface of the S-layer of Bacillus coagulans E38-66/v1. The image was taken under water with a nominal loading force of 100 pN. The surface corrugation corresponding to the gray scale from black to white is 1.8 nm. Bar, 100 nm. (b) Computer image reconstruction of the S-layer shown in (a) obtained after averaging over appr. 500 motifs. The S-layer lattice shows an oblique unit cell (a=8 nm, b=10 nm, base angle = 80°) with lattice symmetry p1. The resolution of molecular details is in the range of 1.5 nm. Bar=10 nm.

Comparative studies of individual strains of many species of eubacteria revealed great diversity regarding lattice types, lattice constants and chemistry of S-layer lattices. In *Aeromonas salmonicida* negatively stained preparations suggested that one type of S-layer protein is even capable of assembling into two distinct structural types on the cell surface.[40,41] (See also chapters 3, 4 and 5). From these data, including homology comparisons of the protein sequences, it became quite obvious that S-layers are non-conservative structures of limited taxonomical value. Similar to other macromolecules constituting the cell surface of prokaryotic cells (e.g. lipopolysaccharides in Gram-negative eubacteria), S-layers frequently only represent a strain-specific characteristic.[15,17,42-48]

2.4. MORPHOGENESIS OF S-LAYERS

Due to the fact that S-layers possess a high degree of structural regularity and that the constituent subunits are the most abundant of all cellular proteins, these crystal lattices are very appealing models for studying the morphogenesis of a supramolecular structure during cell growth.[6,23,25,49,50] Since S-layers are found in Gram-positive and Gram-negative archaeobacteria and eubacteria they can be associated with quite different cell envelope components (Fig. 2.1). Consequently, studies on the morphogenesis of S-layer lattices must also focus on how the constituent tertiary protein subunits assemble on supporting supramolecular structures into two-dimensional arrays during cell growth. An intact "closed" S-layer on an average-sized, rod-shaped cell was calculated to consist of approximately 5×10^5 monomers. Thus, at a generation time of about 20 minutes, at least 500 copies of a single polypeptide species with a molecular weight of approximately 100,000 have to be synthesized, translocated to the cell surface, and incorporated into the S-layer lattice per second.[50] This continuous formation, incorporation and rearrangement of S-layer subunits during cell growth and division can be characterized best as an entropy-driven dynamic process of assembly.[23,51] Structural and chemical studies indicate that in many organisms the rate of synthesis of S-layer subunits is strictly controlled since only minor amounts of S-layer protein can be detected in the growth medium. Only a few organisms have been described which synthesize and shed a significant excess of S-layer protein into the medium.[6,15,52-56] Shedding of S-layer fragments may be also specifically induced by environmental or cultural conditions.[6,54,55,57]

2.4.1. MORPHOGENESIS AND ASSEMBLY IN VITRO

Important information about the mode of morphogenesis of S-layers on intact cells were derived from self-assembly studies of isolated S-layer subunits in vitro.[50] A wide range of methods have been developed for the detachment of S-layers from Gram-positive and Gram-negative eubacteria and archaeobacteria and for disintegration of the lattices into their constituent subunits.[6,15,17,49,58-62] Although considerable differences exist with respect to the mechanical stability and ease of disintegration of S-layer lattices, the subunits of most crystalline arrays interact with each other and with the supporting envelope layers by a combination of ionic bonds involving divalent cations or direct interactions of polar groups, hydrogen bonds and hydrophobic interactions. Most commonly in Gram-positive organisms, a complete disintegration of S-layers into monomers can be obtained by treatment of intact cells or cell walls with high concentrations of hydrogen-bond breaking agents (e.g. urea or guanidinium hydrochloride). Particularly S-layers from Gram-negative bacteria may also disintegrate upon application of metal-chelating agents (e.g. EDTA, EGTA), cation substitution (e.g. Na^+ to replace Ca^{2+}), pH changes (e.g. pH < 4.0) or detergents.[6,15,17,28,50,61-64] A unique extraction procedure was reported for the S-layer of *Lactobacillus helveticus* ATCC 12046. Treatment with 5M lithiumchloride completely removed the S-layer from intact cells without decreasing the cell viability significantly.[65] In certain cases even washing cells with diionized water can lead to dissociation of the S-layer. Repeated extractions with water were shown to be sufficient for isolation of the S-layer protein of intact cells of *Campylobacter fetus*[66] and of *Bacillus thermoaerophilus* DSM 10155 (formerly *B. brevis* ATCC 12990).[67]

From the various extraction and disintegration experiments it can be concluded that the intersubunit bonds in the S-layer are stronger than those binding the crystalline array to the supporting envelope layer. This characteristic is particularly seen as major requirement for recrystallization of S-layers during cell growth.[23,49,64,68]

Some archaeobacterial S-layers were shown to be highly resistant to common denaturing agents, indicating that adjacent subunits may be stabilized even by covalent bonds.[3,11,24,28,69-74]

So far, only a few data are available about specific interactions between S-layers and the underlying supramolecular envelope structures. The most detailed studies have been performed for Gram-negative eubacteria. *Caulobacter* spp. possess S-layer proteins with molecular weights ranging from 100,000 to 193,000 which are immunologically related[75] and assemble into hexagonally ordered S-layer lattices.[76] Attachment of the S-layer subunits occurs to specific oligosaccharide-containing surface molecules, for which Ca^{2+}-ions are required.[77-79] Sequencing of the gene (*RsaA*) encoding the 130,000 molecular weight S-layer protein from *C. crescentus* revealed the existence of a Ca^{2+}-binding region[80] showing homology to other Ca^{2+}-stabilized proteins such as hemolysine. In the case of the (*RsaA*), the Ca^{2+}-binding motif is a nine amino acid repeat, centered with an aspartic residue and is repeated at least four times. As described for other *Caulobacter* spp. as well as in *Campylobacter fetus*, subsp. *fetus*, the N-terminal region of the S-layer protein is responsible for adhesion of the crystalline array to the outer membrane by Ca^{2+}-ions mediated interactions.[81,82] In *Aeromonas hydrophila* TF7 the C-terminal region of the S-layer protein was shown to be essential for anchoring the subunits to the outer membrane whereas the N-terminal part carries the immunodominant surface-exposed regions.[83] For the Gram-negative *Aquaspirillum serpens* MW5 and the Gram-positive *Bacillus brevis* 47 it could be demonstrated that two superimposed S-layers are present which are assembled sequentially. The 150,000 molecular weight S-layer protein from *A. serpens* MW5 formed an array on the outer membrane surface, whereas the 125,000 molecular weight S-layer protein required that array as a template for its in vitro assembly.[63] Polyclonal antiserum raised against the middle S-layer protein from *B. brevis* 47 did not crossreact with the outer S-layer protein,[84] but recognized the S-layer proteins from other *B. brevis* strains which had only one S-layer.[85] This indicated that the S-layer protein directly bound to the peptidoglycan has conserved domains in different *B. brevis* strains. For *Lactobacillus buchneri* it was shown that the isolated S-layer protein could recrystallize to the peptidoglycan-containing rigid cell wall layer in an oblique lattice type identical to that observed on intact cells. Selective extraction of either teichoic acids or neutral sugars from the peptidoglycan-containing sacculi confirmed that a polymer composed of neutral sugars is responsible for the specific adhesion of the S-layer protein to the rigid cell wall layer.[86] Similarly, a cell wall polysaccharide composed of glucosamine was shown to be involved in attachment of the S-layer to the peptidoglycan-containing sacculi of *B. sphaericus* P1.[87] Sequence analysis of the S-layer from *Thermoanaerobacter kivui* (formerly *Acetogenium kivui*)[88] and comparison with data from the S-layer protein of *B. sphaericus*, the middle S-layer protein from *B. brevis* and the hexagonal S-layer protein from *B. brevis* HPD31[89] revealed that the S-layer proteins have homologous sequences (SLH) at their N-terminal regions.[90] On the other hand S-layer proteins not in direct contact with the peptidoglycan-containing layers such as those from *C. crescentus*,[91] *A. salmonicida*[92] and the outer wall protein from *B. brevis*[84] did not show SLH-sequences. Consequently it was suggested that SLH-sequences at the N-terminal region of S-layer proteins could be involved in attachment of the crystalline arrays to the peptidoglycan.[90] Chemical analysis of peptidoglycan-containing sacculi from *B. stearothermophilus* PV72 and its S-layer deficient variant T5 showed that the peptidoglycan belongs to the directly crosslinked meso-DAP-A1 type.[93] A surplus of glucose and glucosamine which were determined to represent about 10% of the dry weight of the peptidoglycan-containing wall layer was attributed to secondary cell wall polymers. Recently it could be confirmed that not the peptidoglycan itself but these ancillary wall polymers are necessary for in vitro crystallization of S-layer subunits onto the rigid cell wall layer (Sára M, Kuen B, Mayer HF et al. Dynamics in oxygen-induced changes in S-layer protein synthesis and cell wall composition in continuous culture from *Bacillus stearothermophilus* PV72 and the S-layer-deficient variant T5. Submitted).

Isolated S-layer subunits from many organisms have been shown to maintain the ability to assemble into two-dimensional arrays in suspension or on suitable surfaces or interfaces upon removal of the disrupting agent used for their isolation.[50,51,58,59,68,77,94-97] (See

Fig. 2.6. Electron micrograph of a negatively stained sheet-like S-layer self-assembly product obtained by recrystallization of isolated S-layer subunits from Clostridium thermosaccharolyticum *D120-70. Lattice symmetry p4. Bar = 100 nm.*

also chapter 8.) Depending on the type of lattice and assembly conditions (e.g. pH, temperature, ionic strength, ion composition), isolated S-layer subunits may recrystallize either into flat sheets (Fig. 2.6), open-ended cylinders or closed vesicles.[49-51] Occasionally, single- and double-layer assembly products are formed, whereby the two constituent monolayers can bind either with their inner face or their outer face. For the halobacterium *Haloferax volcanii* it was shown that detachment and redeposition of the S-layer could be induced by modulating the relative concentration of mono- and divalent cations.[98] Under specific salt conditions, reorganization of the envelope constituents occurred, resulting in the formation of geometrically defined envelope-like structures.[99] All these recrystallization experiments clearly demonstrated that S-layers are entropy-driven self-assembly systems in which all the information for crystallization into regular arrays resides within the individual monomer.[23,50,51] Numerous in vitro recrystallization experiments with isolated S-layer subunits indicate that the formation

of crystalline arrays is initiated by a rapid nucleation of the subunits into oligomeric precursors followed by a subsequent slower
assembly into larger lattice domains.[100-102] (See also chapter 8.) High
resolution electron microscopical studies, labeling experiments with
charged marker molecules, different adsorption assays on solid surfaces, electrostatic and hydrophobic interaction chromatography
revealed that S-layer lattices are anisotropic with regard to their
morphological and physicochemical surface properties. The most
detailed self-assembly studies in suspension and on different supports and matrices have been performed with S-layer subunits
isolated from different *Bacillaceae* including *Bacillus, Clostridium,
Thermoanaerobacter* and *Desulfotomaculum* spp. They have shown
that the inner- and outer surface of these S-layers reveal significant differences in net charge, hydrophobicity, and binding
properties to cell wall polymers.[50,103-108] These characteristics of
S-layer lattices appears to be essential for the proper orientation of
subunits during local insertion in the regular array in the course
of lattice extension during cell growth. (See also chapter 8.)

2.4.2. Extension of S-Layer Lattices on Intact Cells

Numerous studies have been performed to elucidate the dynamic process of assembly of S-layers during cell growth.[50]
Freeze-etching preparations of rod-shaped Gram-positive (Fig. 2.2)
and Gram-negative eubacteria generally reveal a characteristic
orientation of the lattice with respect to the longitudinal axis of
the cylindrical part of the cell.[23] This characteristic feature was
seen as strong indication that S-layers are "dynamic closed surface
crystals" with the intrinsic tendency to assume continuously a
structure of low free energy during cell growth.[49-51] Valuable information about the mechanism involved in the development
and maintenance of crystalline arrays of macromolecules on a growing cell surface also came from reconstitution experiments with
isolated S-layers on cell surfaces from which they had been removed (homologous reattachment) or on those of other
organisms (heterologous reattachment).[49,68] Results of homologous
and heterologous recrystallization experiments of S-layers clearly
demonstrated that the formation of the regular patterns entirely
resides in the subunits themselves and is not affected by the matrix

of the supporting cell envelope layer. Lattices reconstituted on cell envelopes which had maintained their cylindrical shape frequently revealed the orientation with respect to the longitudinal axis of the cell as observed by freeze-etching of intact cells. These results show that the curvature of the cylindrical part of the cell induces an orientation of the lattice with least strains between adjacent subunits. The spherical curvature on cell poles and septation sites or on the whole surface of coccoidal cells allows a random orientation of S-layer lattices.[23,49,51] For maintenance of good long range order of lattices during cell growth S-layer protomers must have the ability to recrystallize on the supporting envelope layer. Labeling experiments with fluorescent antibodies and colloidal gold/antibody marker methods showed that different patterns of S-layer lattice extensions exist for Gram-positive and Gram-negative eubacteria. In Gram-positive eubacteria, lattice growth was shown to occur primarily by insertion of multiple bands or helically arranged bands of S-layers on the cylindrical part of the cell.[50,109,110] In Gram-negative eubacteria, however, insertion of new subunits occurs at random.[111,112] In both types of organisms entirely new S-layer lattices also appear at regions of incipient cell division and the newly formed cell poles. Lattice faults such as dislocations and disclinations frequently observed on S-layers in freeze-etched preparations of intact cells (Fig. 2.2) have been suggested to function as incorporation sites for new subunits.[23,25,50,113]

With the exception of those Gram-negative archaeobacteria in which the S-layer represents the sole envelope component external to the cytoplasmic membrane (Fig. 2.1a), the exported polypeptides or oligomeric precursors forming the crystalline arrays have to pass or even diffuse laterally within intermediate layers (Fig. 2.1c-d) before reaching their incorporation sites. Investigations on mesophilic and thermophilic *Bacillaceae* revealed that a pool of S-layer subunits is present within the network of the peptidoglycan-containing layer.[114] In this context it can be assumed that the translational transport of the S-layer protein is regulated by the packing density of the S-layer subunits within the periplasm delineated by the S-layer and the cytoplasmic membrane.[114,115] Evidence that S-layer subunits must be available over the entire peptidoglycan network in Gram-positive eubacteria came

from investigations on *Lactobacillus helveticus.*[65] On intact cells
from which the S-layer had been stripped by treatment with
5M lithium chloride, new S-layer crystallites appeared all over the
cell surface in a random fashion during exponential phase
growth. Recent studies of *Bacillus stearothermophilus* strains in
continuous culture revealed that depending on growth and environmental conditions individual cells are capable of synthesizing
different S-layer proteins[116,117] (Sára M, Kuen B, Mayer HF et al.
Dynamics in oxygen-induced changes in S-layer protein synthesis
and cell wall composition in continuous culture from *Bacillus
stearothermophilus* PV72 and the S-layer-deficient variant T5. Submitted). During switching in synthesis from S-layer protein
characteristic for the wild-type strain to the S-layer protein of the
variant, patches of different types of S-layer lattices could be
demonstrated on individual cells (Fig. 2.7). This clearly demonstrated that the different S-layer proteins present in the pool
within the peptidoglycan-containing layer specifically incorporate
into corresponding lattices. Upon subsequent growth the new
type of S-layer lattice completely replaced the arrays characteristic
for the wild-type strain. Very few data are available about the
translocation of subunits from the site of synthesis to areas of lattice growth in Gram-negative eubacteria. Two alternative pathways

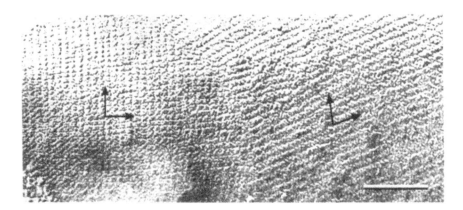

*Fig. 2.7. Electron micrograph of a freeze-etched preparation of intact cells of Bacillus
stearothermophilus NRS 2004/3a. During formation of physiologically induced variants in
continuous culture individual cells are covered with patches of square (p4) and oblique (p2)
S-layer lattices. Upon prolonged cultivation the p2 lattice becomes completely replaced by the
p4 lattice. Bar = 100 nm. (Modified from Sára M, Sleytr UB. J Bacteriol 1994; 176:7182-89.)*

have been considered: one over fusions between the cytoplasmic membrane and the outer membrane[111,118,119] and another through the periplasmic space and the outer membrane, apparently involving an S-layer-specific transport mechanism.[120,121] There is no evidence for pools of S-layer proteins in the cytosol. Labeling experiments on many Gram-positive and Gram-negative eubacteria have demonstrated that once the subunits are incorporated into the regular array they undergo virtually no turnover.

Several archaeobacterial and eubacterial species have been shown to assemble double S-layer lattices or multilayered protein cellular envelopes composed of S-layers and sheaths which closely resemble crystalline cell surface layers (e.g. *Nitrosocystis oceanus*,[122] *Aquaspirillum serpens* MW5,[63,123] *Bacillus brevis* 47,[48,124] *Aquaspirillum sinuosum*,[125] *Lampropedia hyalina*,[126-130] *Pyrobaculum organotrophum* H10,[131] *Methanospirillum hungatei* GP1[132,133] and *Thermoproteus uzoniensis*[134]). Studies on the morphogenesis of such multilayered crystalline arrays have focused on how the constituent tertiary protein subunits specifically assemble into different lattices once emerging on the surface of the protoplast. Such multicomponent self-assembly systems will require different and highly specific bonding properties between the subunits generating the individual lattices. Detailed studies on *A. serpens* have shown that the inner S-layer assembles entropically by itself on the surface of the outer membrane whereas the outer S-layer requires the inner layer as template for assembly.[63]

An even more complex multicomponent assembly system composed of regular arrays was described for *Methanospirillum hungatei*. This organism assembles external to the plasma membrane a hexagonally ordered S-layer composed of two polypeptides[37] and a very stable sheath with heteropolymer polypeptide composition and oblique lattices.[135,136] In addition, individual cells within the filamentous sheath are separated by paracrystalline multilayered spacer-plugs composed of different protein species.[137] Unlike Gram-positive eubacteria and archaeobacteria (Fig. 2.2c) and Gram-negative eubacteria (Fig. 2.2d), where S-layers are associated with other cell wall components, in most archaeobacteria (Fig. 2.1a-b) the dynamic process of assembly of S-layers occurs directly on the surface of the cytoplasmic membrane.

Evidence for the involvement of an S-layer in cell morphology and division was reported for the Gram-negative archaeobacterium *Methanocorpusculum sinense*.[25] Cells of this organism reveal, like to many other archaeobacteria which have an S-layer as their sole cell wall constituent, a highly lobed cell structure with a hexagonally arranged S-layer. Although, lattice faults such as dislocations and disclinations are a geometrical necessity on the surface of a closed protein crystal,[23,50,113,138] detailed analyses of high resolution freeze-etching replicas of intact cells supported the notion that they also play important roles as sites for the incorporation of new S-layer subunits, in the formation of the lobed structure, and in the cell division process. The latter was shown to be determined by the ratio between the increase of protoplast volume and the increase in actual S-layer surface area during cell growth. It was proposed that this relatively simple regulation mechanism could represent a common feature in lobed Gram-negative archaeobacteria.[25] (See also chapter 5.)

Further evidence for the mode of lattice growth and the shape determining function of an S-layer in a Gram-negative archaeo-bacteria was derived from analyses of lattice faults in the surface array of *Thermoproteus tenax* and *Thermoproteus neutrophilus*.[73,139] Both organisms are rod-shaped and covered with a hexagonally arranged S-layer lattice. The lattice shows no defects at the cylindrical part of the cell but a defined number of wedge disclinations at the hemispherical cell poles. These morphological data strongly support the notion that the radius of curvature of the lattice and, therefore, the diameter of the cells would be determined by the mass distribution and bonding properties of the individual protomeric units. Elongation of the cylindrical part of the cell most probably involves incorporation of protomeric subunits at sites of sliding dislocations at the cell poles.[73]

2.5. CONCLUSIONS

Four decades after the first observation of an S-layer[140] two-dimensional crystalline arrays of protein or glycoprotein subunits are recognized as surface structures common amongst species of nearly every taxonomical group of walled eubacteria and as an almost universal feature of archaeobacterial envelopes. The assembly and recrystallization properties of S-layer protomers guarantee the maintenance of a closed isoporous cell surface layer with a low number of faults in the array during cell growth and division. The only necessity for maintaining such a crystalline surface layer is the synthesis of an excess of protein or glycoprotein subunits and their transfer to the cell surface. Theoretically, there is no other possibility of forming such a highly ordered, closed protein membrane on a growing cell surface with less redundancy of information. The entropy-driven morphogenesis of an S-layer only requires the genetic information for a single polypeptide chain. The surface and permeability (barrier) properties of S-layers are only determined by the mass distribution and physicochemical surface properties of the protomers composing the lattice. Due to their anisotropic surface properties S-layer subunits can form well-defined specific associations with other supramolecular cell envelope components including lipid membranes (see also chapter 8). Consequently, it is tempting to speculate, that S-layer like primitive "closed protein membranes" with the ability for dynamic growth could have fulfilled all necessary morphogenetic and barrier functions required by a self-reproducing system during the early stages of biological evolution.[51]

The high organization of parameters of S-layers, particularly their surface location, self assembly capabilities and their abundance as the dominant cellular protein make them an exciting supramolecular structure for basic molecular biological studies (see chapters 3, 4 and 5) and very diverse fields of applied research (see chapters 6, 7 and 8).

Acknowledgment

This work was supported in part by the Austrian Science Foundation, Project S72, and by the Austrian Federal Ministry for Science, Research and the Arts.

References

1. Sleytr UB, Glauert AM. Bacterial cell walls and membranes. In: Harris JR, ed. Electron Microscopy of Proteins. Vol 3. London: Academic Press, 1982:41-76.
2. Beveridge TJ. A fundamental design of bacteria. In: Poindexter JS, Leadbetter ER, eds. Bacteria in Nature, Structure, Physiology, and Genetic Adaptability. Vol 3. New York: Plenum, 1989:1-65.

3. Beveridge TJ, Graham LL. Surface layers of bacteria. Microbiol Rev 1991; 55:684-705.

4. Sleytr UB, Messner P. Crystalline bacterial cell surface layers (S-layers). In: Lederberg J, ed. Encyclopedia of Microbiology. Vol 1. San Diego: Academic Press, 1992:605-14.

5. Hancock IC, Poxton IR, eds. Bacterial Cell Surface Techniques. Chichester: Wiley, 1988.

6. Sleytr UB. Regular arrays of macromolecules on bacterial cell walls: structure, chemistry, assembly and function. Int Rev Cytol 1978; 53:1-64.

7. Sleytr UB, Messner P. Crystalline surface layers on bacteria. In: Sleytr UB, Messner P, Pum D, Sára M, eds. Crystalline Bacterial Cell Surface Layer. Berlin: Springer, 1988:160-86.

8. Roberts K, Grief C, Hills GJ et al. Cell wall glyocoproteins: structure and function. J Cell Sci 1985; 2(Suppl.):105-27.

9. Walker PD, Short JA, Roper G et al. The structure of clostridial spores. In: Fuller RF, Lovelock DW, eds. Microbial Ultrastructure. London: Academic Press, 1976:117-46.

10. Kandler O, König H. Cell envelopes of archaebacteria. In: Woese CR, Wolfe RS, eds. The Bacteria, Archaebacteria. Vol VIII. New York: Academic Press, 1985:413-57.

11. Kandler O, König H. Cell envelopes of archaea: structure and chemistry. In: Kates N, Kushner DJ, Matheson AT, eds. The Biochemistry of Archaea (Archaebacteria). Amsterdam: Elsevier, 1993:223-59.

12. Sleytr UB, Messner P, Sára M et al. Crystalline envelope layers in archaebacteria. System Appl Microbiol 1986; 7:310-13.

13. Woese CR, Kandler O, Wheelis ML. Towards a natural system of organisms: proposal for the domains Archaea, Bacteria, and Eucarya. Proc Natl Acad Sci USA 1990; 87:4576-79.

14. Kandler O. Cell wall biochemistry and three-domain concept of life. System Appl Microbiol 1994; 16:501-9.

15. Sleytr UB, Messner P. Crystalline surface layers on bacteria. Annu Rev Microbiol 1983; 37:311-39.

16. Beveridge TJ. Mechanism of Gram-variability in selected bacteria. J Bacteriol 1990; 172:1609-20.

17. Messner P, Sleytr UB. Crystalline bacterial cell surface layers. In: Rose AH, ed. Advances in Microbial Physiology. Vol 33. London: Academic Press, 1992:213-75.

18. Messner P, Sleytr UB. Bacterial surface layer glycoproteins. Glycobiology 1991; 1:545-51.

19. Messner P., Schuster-Kolbe J, Schäffer C et al. Glycoprotein nature of selected bacterial S-layers. In: Beveridge TJ, Koval SF, eds. Advances in Bacterial Paracrystalline Surface Layers. New York: Plenum, 1993:95-107.

20. Sleytr UB, Messner P, Pum D. Analysis of crystalline bacterial surface layers by freeze-etching, metal shadowing, negative staining and ultrathin sectioning. In: Mayer F, ed. Methods of Microbiology. Vol 20. London: Academic Press, 1988:29-60.

21. Sleytr UB. Gefrierätzung verschiedener Stämme von *Bacillus sphaericus*. Arch Microbiol 1970; 72:238-51.

22. Sleytr UB, Thornley MJ. Freeze-etching of the cell envelope of an *Acinetobacter* species which carries a regular array of surface subunits. J Bacteriol 1973; 116:1383-97.

23. Sleytr UB, Glauert AM. Analysis of regular arrays of subunits on bacterial surfaces; evidence for a dynamic process of assembly. J Ultrastruct Res 1975; 50:103-16.

24. König H. Archaebacterial cell envelopes. Can J Microbiol 1988; 34:395-406.

25. Pum D, Messner P, Sleytr UB. Role of the S-layer in morphogenesis and cell division of the archaeobacterium *Methanocorpusculum sinense*. J Bacteriol 1991; 173:6865-73.

26. Hovmöller S, Sjögren A, Wang DN. The structure of crystalline bacterial surface layers. Prog Biophys Mol Biol 1988; 51:131-63.

27. Baumeister W, Wildhaber I, Phipps BM. Principles of organization in eubacterial and archaebacterial surface proteins. Can J Microbiol 1989; 35:215-27.

28. Beveridge TJ. Bacterial S-layers. Curr Opinion Struct Biol 1994; 4:204-12.

29. Baumeister W, Lembcke G, Dürr R et al. Electron crystallography of bacterial surface proteins. In: Fryer JR, Dorset DL, eds. Electron Crystallography of Organic Molecules. Dordrecht: Kluwer, 1991:283-296.

30. Hovmöller S. Crystallographic image processing applications for S-layers. In: Beveridge TJ, Koval SF, eds. Advances in Bacterial Paracrystalline Surface Layers. New York: Plenum, 1993:13-21.

31. Nußer E, König H. S-layer studies on three species of *Methanococcus* living at different temperatures. Can J Microbiol 1987; 33:256-61.

32. Takumi K, Koga T, Oka T et al. Self-assembly, adhesion and chemical properties of tetragonally arrayed S-layer proteins of *Clostridium*. J Gen Appl Microbiol 1991; 37:455-65.

33. Kawata T, Takeoka A, Takumi K et al. Demonstration and preliminary characterization of a regular array in the cell wall of *Clostridium difficile*. FEMS Microbiol Lett 1984; 24:323-28.

34. Takeoka A, Takumi K, Koga T et al. Purification and characterization of S-layer proteins from *Clostridium difficile* GAI 0714. J Gen Microbiol 1991; 137:261-67.

35. Hagiya H, Oka T, Tsuji H et al. The S-layer composed of two different protein subunits from *Clostridium difficile* GAI 1152: a simple purification method and characterization. J Gen Appl Microbiol 1992; 38:63-74.

36. Peters J, Nitsch M, Kühlmorgen B et al. Tetrabrachion: a filamentous archaebacterial surface protein assembly of unusual structure and extreme stability. J Mol Biol 1995; 245:385-401.

37. Firtel M, Southam G, Harauz G et al. Characterization of the cell wall of the sheathed methanogen *Methanospirillum hungatei* GP1 as an S-layer. J Bacteriol 1993; 175:7550-60.

38. Southam G, Firtel M, Blackford BL et al. Transmission electron microscopy, scanning tunneling microscopy, and atomic force microscopy of the cell envelope layers of the archaeobacterium *Methanospirillum hungatei* GP1. J Bacteriol 1993; 175:1946-55.

39. Firtel M, Xu W, Southam G et al. Tip-induced displacement and imaging of a multilayered bacterial structure by scanning tunneling microscopy. Ultramicroscopy 1994; 55:113-19.

40. Stewart M, Beveridge TJ, Trust TJ. Two patterns in the *Aeromonas salmonicida* A-layer may reflect a structural transformation that alters permeability. J Bacteriol 1986; 166:120-27.

41. Kay WW, Thornton JC, Garduno RA. Structure-function aspects of the *Aeromonas salmonicida* S-layer. In: Beveridge TJ, Koval S, eds. Advances in Bacterial Paracrystalline Surface Layers. New York: Plenum, 1993:151-58.

42. Koval SF. Paracrystalline protein surface arrays on bacteria. Can J Microbiol 1988; 34:407-14.

43. Kay WW, Phipps BM, Ishiguro EE et al. Surface layer virulence A-protein from *Aeromonas salmonicida* strains. Can J Biochem Cell Biol 1984; 62:1064-71.

44. Messner P, Hollaus F, Sleytr UB. Paracrystalline cell wall surface layers of different *Bacillus stearothermophilus* strains. Int J Syst Bacteriol 1984; 34:202-10.

45. Sleytr UB, Sára M, Küpcü Z, et al. Structural and chemical characterization of S-layers of selected strains of *Bacillus stearothermophilus* and *Desulfotomaculum nigrificans*. Arch Microbiol 1986; 146:19-24.

46. Sára M, Sleytr UB. Molecular sieving through S-layers of *Bacillus stearothermophilus* strains. J Bacteriol 1987; 169:4092-98.

47. Lewis LO, Yousten AA, Murray RGE. Characterization of the surface protein layers of the mosquito-pathogenic strains of *Bacillus sphaericus*. J Bacteriol 1987; 169:72-79.

48. Sára M, Moser-Thier K, Kainz U et al. Characterization of S-layers from mesophilic bacillaceae and studies on their protective role towards muramidases. Arch Microbiol 1990; 153:209-14.

49. Sleytr UB. Morphopoietic and functional aspects of regular protein membranes present on prokaryotic cell walls. In: Kiermayer O, ed., Cytomorphogenesis in Plants, Cell Biology Monographs, Vol 8. Wien: Springer, 1981:3-26.

50. Sleytr UB, Messner P. Self-assemblies of crystalline bacterial cell surface layers. In: Plattner H, ed. Electron Microscopy of Subcellular Dynamics. Boca Raton: CRC Press, 1989:13-31.

51. Sleytr UB, Plohberger R. The dynamic process of assembly of two-dimensional arrays of macromolecules on bacterial cell walls. In: Baumeister W, Vogell W, eds. Electron Microscopy at Molecular Dimensions. Berlin: Springer, 1980:36-47.

52. Beveridge TJ. Ultrastructure, chemistry, and function of the bacterial wall. Int Rev Cytol 1981; 72:229-317.

53. Tsukagoshi N, Tabata R, Takemura T et al. Molecular cloning of a major cell wall protein gene from protein-producing *Bacillus brevis* 47 and its expression in *Escherichia coli* and *Bacillus subtilis*. J Bacteriol 1984; 158:1054-60.

54. Luckevich MD, Beveridge TJ. Characterization of a dynamic S-layer on *Bacillus thuringiensis*. J Bacteriol 1989; 171:6656-67.

55. Schultze-Lam S, Harauz G, Beveridge TJ. Participation of a cyanobacterial S-layer in fine-grain mineral formation. J Bacteriol 1992; 174:7971-81.

56. Farchaus JW, Ribot WJ, Dows MB et al. Purification and characterization of the major surface array protein form the avirulent *Bacillus anthracis* Delta Sterne-1. J Bacteriol 1995; 177:2481-89.

57. Sleytr UB, Glauert AM. Ultrastructure of the cell walls of two closely related *Clostridia* that possess different regular arrays of surface subunits. J Bacteriol 1976; 126:869-82.

58. Sleytr UB. Self-assembly of the hexagonally and tetragonally arranged subunits of bacterial surface layers and their reattachment to cell walls. J Ultrastruct Res 1976; 55:360-77.

59. Beveridge TJ, Murray RGE. Reassembly in vitro of the superficial cell wall components of *Spirillum putridiconchylium*. J Ultrastruct Res 1976; 55:105-18.

60. Hastie AT, Brinton Jr CC. Isolation, characterization and in vitro self-assembly of the tretragonally arranged layer of *Bacillus sphaericus*. J Bacteriol 1979; 138:999-1009.

61. Koval SF, Murray RGE. The isolation of surface array proteins from bacteria. Can J Biochem Cell Biol 1984; 62:1181-89.

62. Messner P, Sleytr UB. Separation and purification of S-layers from Gram-positive and Gram-negative bacteria. In: Hancock IC, Poxton IR, eds. Bacterial Cell Surface Techniques. Chichester: Wiley, 1988:97-104.

63. Kist ML, Murray RGE. Components of the regular surface array of *Aquaspirillum serpens* MW5 and their assembly in vitro. J Bacteriol 1984; 157:599-606.

64. Bayley DP, Koval SF. Membrane association and isolation of the S-layer protein of *Methanoculleus marisnigri*. Can J Microbiol 1993; 40:237-41.

65. Lortal S, van Heijenoort J, Gruber K et al. S-layer of *Lactobacillus helveticus* ATCC 12046: isolation, chemical characterization and reformation after extraction with lithium chloride. J Gen Microbiol 1992; 138:611-18.

66. Pei Z, Ellison III RT, Lewis RV et al. Purification and characterization of a family of high molecular weight surface-array proteins from *Campylobacter fetus*. J Biol Chem 1988; 263:6414-20.

67. Kosma P, Wugeditsch T, Christian R et al. Glycan structure of a heptose-containing S-layer glycoprotein of *Bacillus thermoaerophilus*. Glycobiology 1995; in press.

68. Sleytr UB. Heterologous reattachment of regular arrays of glycoproteins on bacterial surfaces. Nature 1975; 257:400-2.

69. Beveridge TJ, Stewart M, Doyle RJ et al. Unusual stability of the *Methanospirillum hungatei* sheath. J Bacteriol 1985; 162:728-37.

70. Beveridge TJ, Harris BJ, Patel GB et al. Cell division and filament splitting in *Methanothrix concilii*. Can J Microbiol 1986; 32:779-86.

71. Beveridge TJ, Harris BJ, Sprott GD. Septation and filament splitting in *Methanospirillum hungatei*. Can J Microbiol 1987; 33:725-32.

72. Beveridge TJ, Sára M, Pum D et al. The structure, chemistry and physicochemistry of the *Methanospirillum hungatei* GP1 sheath. In: Sleytr UB, Messner P, Pum D et al, eds. Crystalline Bacterial Cell Surface Layers. Berlin: Springer, 1988:26-30.

73. Messner P, Pum D, Sára M et al. Ultrastructure of the cell envelope of the archaebacteria *Thermoproteus tenax* and *Thermoproteus neutrophilus*. J Bacteriol 1986; 166:1046-54.

74. Kandler O. Comparative chemistry of the rigid cell wall component and its phylogenetic implications. In: Sleytr UB, Messner P, Pum D et al, eds. Crystalline Bacterial Cell Surface Layers. Berlin: Springer, 1988:1-6.

75. Walker SG, Smith SH, Smit J. Isolation and comparison of the paracrystalline surface layer proteins of freshwater Caulobacters. J Bacteriol 1992; 174:1783-92.

76. Smit J, Engelhardt H, Volker S et al. The S-layer of *Caulobacter crescentus:* three-dimensional image reconstruction and structural analysis by electron microscopy. J Bacteriol 1992; 174:6527-38.

77. Bingle WH, Walker SG, Smit J. Definition of form and function for the S-layer of *Caulobacter crescentus*. In: Beveridge TJ, Koval SF, eds. Advances in Bacterial Paracrystalline Surface Layers. New York: Plenum, 1993:181-93.

78. Walker SG, Smit J. Attachment of the S-layer of *Caulobacter crescentus* to the cell surface. In: Beveridge TJ, Koval SF, eds. Advances in Paracrystalline Cell Surface Layers. New York: Plenum, 1993:289-91.

79. Walker SG, Nedra-Karunaratne D, Ravenscroft N. et al. Characterization of mutants of *Caulobacter crescentus* defective in surface attachment of the paracrystalline surface layer. J Bacteriol 1994; 176:6312-23.

80. Gilchrist A, Fisher JA, Smit J. Nucleotide sequence analysis of the gene encoding the *Caulobacter crescentus* paracrystalline surface layer protein. Can J Microbiol 1992; 38:193-202.

81. Dworkin J, Tummuru MKR, Blaser MJ. A lipopolysaccharide binding domain of *Campylobacter fetus* S-layer protein resides within the conserved N-terminus of a family of silent and divergent genes. J Bacteriol 1995; 177:1734-41.

82. Yang L, Pei Z, Fujimoto S et al. Reattachment of surface array proteins to *Campylobacter fetus* cells. J Bacteriol 1992; 174:1258-67.

83. Kostrzynska M, Dooley JSG, Shimojo T et al. Antigenic diversity of the S-layer proteins from pathogenic strains of *Aeromonas hydrophila* and *Aeromonas veronii* biotype sobria. J Bacteriol 1992; 174:40-47.

84. Tsuboi A, Uchihi R, Takahashi Y et al. Characterization of the genes coding two major cell wall proteins from protein-producing *Bacillus brevis* 47. J Bacteriol 1986; 168:365-73.

85. Gruber K, Tanahashi H, Tsuboi A et al. Comparative study on the cell wall structure of protein-producing *Bacillus brevis*. FEMS Lett 1988; 56:113-18.

86. Masuda K, Kawata T. Characterization of a regular array in the cell wall of *Lactobacillus buchneri* and its reattachment to the other wall components. J Gen Microbiol 1981; 124:81-90.

87. Hastie AT, Brinton Jr CC. Specific interaction of the tetragonally arrayed protein layer of *Bacillus sphaericus* with the peptidoglycan sacculus. J Bacteriol 1979; 138:1010-21.

88. Peters J, Peters M, Lottspeich F et al. S-layer protein gene of *Acetogenium kivui*: cloning and expression in *Escherichia coli* and determination of the nucleotide sequence. J Bacteriol 1989; 171:6307-15.

89. Ebisu S, Tsuboi A, Takagi H et al. Conserved structures of cell wall genes among protein producing *Bacillus brevis* strains. J Bacteriol 1990; 172:1312-20.

90. Lupas A, Engelhardt H, Peters J et al. Domain structure of the *Acetogenium kivui* surface layer revealed by electron crystallography and sequence analysis. J Bacteriol 1994; 176:1224-33.

91. Smit J, Agabian N. Cloning of the major protein of *Caulobacter crescentus* periodic surface layer: deletion and characterization of the cloned peptide by protein expression analysis. J Bacteriol 1984; 160:1137-45.

92. Chu S, Cavaignac S, Feutrier J et al. Structure of the tetragonal surface virulence array protein and gene of *Aeromonas salmonicida.* J Biol Chem 1991; 266:15258-65.

93. Schuster C, Mayer H, Kieweg R et al. A synthetic medium for continuous culture of the S-layer carrying *Bacillus stearothermophilus* PV72 and studies on the influence of growth conditions on cell wall properties. Biotechnol Bioeng 1995; 48:66-77.

94. Beveridge TJ, Murray RGE. Superficial cell wall layers on *Spirillum* "Ordal" and their in vitro reassembly. Can J Microbiol 1976; 22:567-82.

95. Buckmire FLA, Murray RGE. Substructure and in vitro assembly of the outer structural layer of *Spirillum serpens.* J Bacteriol 1976; 125:290-99.

96. Koval SF, Murray RGE. Effect of calcium on the in vitro assembly of the surface protein of *Aquaspirillum serpens* VHA. Can J Microbiol 1985; 31:261-67.

97. Messner P, Pum D, Sleytr UB. Characterization of the ultrastructure and the self-assembly of the surface layer of *Bacillus stearothermophilus* strain NRS 2004/3a. J Ultrastruct Mol Struct Res 1986; 97:73-88.

98. Cohen S, Kessel M, Shilo, M. Nature of the salt dependence of the envelope of a Dead Sea archaebacterium, *Haloferax volcanii.* Arch Microbiol 1991; 156:198-203.

99. Cohen S, Shilo M, Kessel M. In vitro spontaneous reorganization of *Haloferax volcanii* envelope material into geometrical forms. Arch Microbiol 1993; 160:248-52.

100. Jaenicke R, Welsch R, Sára M et al. Stability and self-assembly of the S-layer protein of the cell wall of *Bacillus stearothermophilus.* Biol Chem Hoppe-Seyler 1985; 366:663-70.

101. Pum D, Weinhandl M, Hödl C et al. Large-scale recrystallization of the S-layer of ·*Bacillus coagulans* E38-66 at the air/water interface and on lipid films. J Bacteriol 1993; 175:2762-66.

102. Pum D, Sleytr UB. Large-scale reconstruction of crystalline bacterial surface layer proteins at the air-water interface and on lipids. Thin Solid Films 1994; 244:882-86.

103. Sleytr UB, Messner P. Crystalline surface layers in prokaryotes. J Bacteriol 1988; 170:2891-97.

104. Sára M, Sleytr UB. Charge distribution on the S-layer of *Bacillus stearothermophilus* NRS 1536/3c and importance of charged groups for morphogenesis and function. J Bacteriol 1987; 169:2804-9.

105. Sára M, Sleytr UB. Relevance of charged groups for the integrity of the S-layer from *Bacillus coagulans* E38-66 and for molecular interactions. J Bacteriol 1993; 175:2248-54.

106. Pum D, Sára M, Sleytr UB. Structure, surface charge, and self-assembly of the S-layer lattice from *Bacillus coagulans* E38-66. J Bacteriol 1989; 171:5296.

107. Gruber K, Sleytr UB. Influence of an S-layer on surface properties of *Bacillus stearothermophilus*. Arch Microbiol 1991; 156:181-85.

108. Sára M, Pum D, Sleytr UB. Permeability and charge-dependent adsorption properties of the S-layer lattice from *Bacillus coagulans* E38-66. J Bacteriol 1992; 174:3487-93.

109. Howard LV, Dalton DD, McCoubrey Jr WK. Expansion of the tetragonally arrayed cell wall protein layer during growth of *Bacillus sphaericus*. J Bacteriol 1982; 149:748-57.

110. Gruber K, Sleytr UB. Localized insertion of new S-layer during cell growth of *Bacillus stearothermophilus* strains. Arch Microbiol 1988; 149:485-91.

111. Smit J. Protein surface layers of bacteria. In: Inouye M, ed. Bacterial Outer Membranes as Model Systems. New York: Wiley, 1987:343-76.

112. Smit J, Agabian N. Cell surface patterning and morphogenesis: biogenesis of a periodic surface array during *Caulobacter* development. J Cell Biol 1982; 95:41-49.

113. Harris WF. Disclinations. Sci Am 1977; 237:130-45.

114. Breitwieser A, Gruber K, Sleytr UB. Evidence for an S-layer protein pool in the peptidoglycan of *Bacillus stearothermophilus*. J Bacteriol 1992; 174:8008-15.

115. Beveridge TJ. The periplasmic space and the periplasm in Gram-positive and Gram-negative bacteria. ASM News 1995; 61:125-30.

116. Sára M, Sleytr UB. Comparative studies of S-layer proteins from *Bacillus stearothermophilus* strains expressed during growth in continuous culture under oxygen-limited and non-oxygen-limited conditions. J Bacteriol 1994; 176:7182-89.

117. Sára M, Pum D, Küpcü S et al. Isolation of two physiologically induced variant strains of *Bacillus stearothermophilus* NRS 2004/3a and characterization of their S-layer lattices. J Bacteriol 1994; 176:848-60.

118. Bayer ME. Ultrastructure and organization of the bacterial envelope. Ann New York Acad Sci 1974; 235:6.

119. Smit J, Nikaido H. Outer membrane of Gram-negative bacteria. XVIII. Electron microscopical studies on porin insertion sites and growth of cell surface of *Salmonella typhimurium*. J Bacteriol 1978; 135:687-702.

120. Belland RJ, Trust TJ. Synthesis, export, and assembly of *Aeromonas salmonicida* A-layer analysed by transposon mutagenesis. J Bacteriol 1985; 163:877-81.

121. Noonan B, Trust TJ. Molecular analysis of an A-protein secretion mutant of *Aeromonas salmonicida* reveals a surface layer-specific protein secretion pathway. J Mol Biol 1995; 248:316-27.

122. Watson SW, Remsen CC. Cell envelope of *Nitrosocystis oceanus*. J Ultrastruct Res 1970; 33:148-60.

123. Stewart M, Murray RGE. Structure of the regular surface layer of *Aquaspirillum serpens* MW5. J Bacteriol 1982; 150:348-57.

124. Tsuboi A, Tsukagoshi N, Udaka S. Reassembly in vitro of hexagonal surface arrays in a protein-producing bacterium, *Bacillus brevis* 47. J Bacteriol 1982; 151:1485-97.

125. Smit SH, Murray RGE. The structure and associations of the double S-layer on the cell wall of *Aquaspirillum sinuosum*. Can J Microbiol 1990; 36:327-35.

126. Pangborn J, Starr MP. Ultrastructure of *Lampropedia hyalina*. J Bacteriol 1966; 91:2025-30.

127. Austin JW, Murray RGE. The perforate component of the regulary structured (RS) layer of *Lampropedia hyalina*. Can J Microbiol 1987; 33:1039-45.

128. Austin JW, Murray RGE. The surface layers of *Lampropedia hyalina*. In: Sleytr UB, Messner P, Pum D et al, eds. Crystalline Bacterial Cell Surface Layers. Berlin: Springer, 1988:17-20.

129. Austin JW, Murray RGE. Isolation and in vitro assembly of the components of the outer S-layer of *Lampropedia hyalina*. J Bacteriol 1990; 172:3681-89.

130. Austin JW, Engel A, Murray RGE et al. Structural analysis of the S-layer of *Lampropedia hyalina*. J Ultrastruct Mol Struct Res 1989; 102:255-64.

131. Phipps BM, Huber R , Baumeister W. The cell envelope of the hyperthermophilic archaebacterium *Pyrobaculum organotrophum* consists of two regularly arrayed protein layers: three-dimensional structure of the outer layer. Mol Microbiol 1991; 5:253-65.

132. Beveridge TJ, Sprott GD, Whippey P. Ultrastructure, inferred porosity, and Gram-staining character of *Methanospirillum hungatei* filament termini describe a unique cell permeability for this archaeobacterium. J Bacteriol 1991; 173:130-40.

133. Southam G, Beveridge TJ. Paracrystalline layers of *Methanospirillum hungatei* GP1. In: Beveridge TJ, Koval SF, eds. Advances in Paracrystalline Cell Surface Layers. New York: Plenum, 1993:129-42.

134. Bonch-Osmolovskaya EA, Miroshnichenko ML, Kostrikina NA et al. *Thermoproteus uzoniensis* sp. nov., a new extremely thermophilic archaebacterium from Kamchatka continental hot springs. Arch Microbiol 1990; 154:556-59.

135. Sprott GD, Beveridge TJ, Patel GB et al. Sheath disassembly in *Methanospirillum hungatei* strain GP1. Can J Microbiol 1986; 32:847-54.

136. Southam G, Beveridge TJ. Detection of growth sites in and promoter pools for the sheath of *Methanospirillum hungatei* GP1 by use of constituent organosulfur and immunogold labelling. J Bacteriol 1992; 174:6460-70.

137. Firtel M, Southam G, Harauz G et al. The organization of the paracrystalline multilayered spacer-plugs of *Methanospirillum hungatei*. J Struct Biol 1994; 112:160-71.

138. Caspar DLD, Klug A. Physical principles in the construction of regular viruses. Cold Spring Harbor Symposium on Quantitative Biology 1962; 27:1-23.

139. Wildhaber I, Baumeister W. The cell envelope of *Thermoproteus tenax*: three-dimensional structure of the surface layer and its role in shape maintenance. EMBO J 1987; 6:1475-80.

140. Houwink AL. A macromolecular monolayer in the cell wall of *Spirillum* spec. Biochim Biophys Acta 1953; 10:360-66.

141. Sleytr UB, Messner P, Pum D et al. Crystalline bacterial cell surface layer: general principles and application potential. J Appl Bacteriol 1993; 74:21S-32S.

CHAPTER 3

CHEMICAL COMPOSITION AND BIOSYNTHESIS OF S-LAYERS

Paul Messner

3.1. INTRODUCTION

The aim of this chapter is to summarize particularly those data on chemistry and biosynthesis of S-layers which have been published since our last comprehensive review on bacterial cell surface layers.[1] It should provide basic information about S-layers for a better understanding of their biosynthesis and possible biological function(s) and for the assessment of the application potential of S-layers in biotechnology.[2] For several S-layer proteins detailed chemical characterizations have been published during the past decade (for reviews see refs. 1, 3-5). An extension of the existing list[1] of S-layer-carrying microorganisms covering new descriptions is given in the appendix. With the advent of molecular biology and genetics it proved practical to discuss results of cloning and sequencing experiments on S-layers in that context. Thus, for those S-layers which have been cloned and sequenced detailed descriptions of their amino acid compositions and structures are provided in chapter 4.

The most frequently applied methods for initial chemical analysis of S-layers of archaeobacteria and eubacteria include sodium dodecyl sulfate-polyacrylamide gel electrophoresis (SDS-PAGE) and

Crystalline Bacterial Cell Surface Proteins, edited by Uwe B. Sleytr, Paul Messner, Dietmar Pum, Margit Sára. © 1996 R.G. Landes Company.

amino acid analysis.[1-6] From SDS-PAGE relevant information about the apparent molecular mass of the constituting SDS-soluble S-layer protomers, their purity and homogeneity and, if relevant, their degree of glycosylation, can be derived. It has been demonstrated that most S-layers are composed of single, high-molecular-weight polypeptide species with apparent molecular masses of 40 to 220 kDa (for review see refs. 1, 3-5). However, there is also evidence that some bacteria are covered with more than one chemically and immunologically not identical S-layer (e.g. *Aquaspirillum serpens,*[7] *Lampropedia hyalina,*[8] *Bacillus brevis,*[9,10] *Clostridium difficile,*[11] *Pyrobaculum organotrophum,*[12] or *Thermococcus stetteri*[13]). For certain select S-layers the macromolecular assembly of a single protomer can be composed of several different subunits. This was elaborated on tetrabrachion, the S-layer-like protein complex of the archaeobacterium *Staphylothermus marinus.*[14] Most chemical analyses have been sustained by electron microscopical examination of freeze-etched or thin-sectioned intact cells or cell wall preparations (see chapter 2).

Amino acid analyses of highly purified S-layers showed that most of these macromolecules possess a rather similar overall composition, independent on the phylogeny of the corresponding organisms.[3,4] Usually S-layers comprise weakly acidic proteins with a 40% to 60% proportion of hydrophobic amino acids and a rather low amount of sulfur-containing amino acids. The isoelectric points of many intact S-layer proteins range from approximately 4 to 6. However, for S-layers of some lactobacilli pI values > 9 have been calculated.[15,16]

Predictions about the secondary structure of different S-layer proteins (e.g. *Aeromonas salmonicida, Campylobacter fetus*) have been derived from comparisons of circular dichroism spectra under native and denaturing conditions.[17,18] Molecular biology has provided information about amino acid composition and conformation of S-layers (see chapter 4 of this volume) which substantiate the results of the chemical analyses. From circular dichroism measurements and sequencing experiments of several S-layers a general picture about their secondary conformation can be drawn. Typically the β-sheet content is in the range of 40% and the β-helix content is in the range of 20%. Aperiodic foldings and β-turn content, however, vary tremendously between 5% and 45% in specific S-layer proteins.[1,17,18]

Since our last review[1] considerable progress in a comprehensive chemical characterization of specific archaeobacterial S-layers (e.g. *Methanospirillum hungatei*,[19,20] *Halobacterium halobium* and *Haloferax volcanii*,[21] *Methanothermus fervidus*[22]) and eubacterial S-layers (e.g. *Aeromonas salmonicida* and *Aeromonas hydrophila*,[23,24] *Campylobacter fetus*,[25] *Caulobacter crescentus*,[26] *Thermoanaerobacter thermohydrosulfuricus*,[27] *Paenibacillus* [formerly *Bacillus*] *alvei*[28]) has been made. Additionally, in the following sections a distinction will be made between non-glycosylated and glycosylated S-layer proteins. This appears to be reasonable since glycosylation is an important post-translational modification of the S-layer protein and most archaeobacterial S-layers are indeed glycoproteins (for review see ref. 29). So far, eubacterial S-layer glycoproteins have only been demonstrated unambiguously on several *Bacillaceae* (for review see refs. 30 and 31).

3.2. NON-GLYCOSYLATED S-LAYER PROTEINS

Not all investigated bacteria have been analyzed in great detail for the presence of covalently linked carbohydrates in their S-layer proteins. Usually glycosylation is demonstrated initially on SDS-polyacrylamide gels either upon periodate oxidation by periodic acid-Schiff staining reaction, or by staining of the gels with Alcian blue or thymol-sulfuric acid, or after blotting and subsequent labeling with specific markers (e.g. digoxygenin). However, for a conclusive assignment, detailed chemical analyses are required. Therefore it might happen that organisms which are listed in this section actually possess glycosylated S-layer proteins.

3.2.1. ARCHAEOBACTERIA

Comparison of cell wall and cell envelope polymers from archaeobacteria have revealed that most organisms are covered by glycosylated S-layer proteins.[29] However, a few non-glycosylated S-layers have been described among the methanogens. They represent the largest group within the archaeobacteria and show some structural and chemical similarities with cell wall components of eubacteria and even eukaryotes.[32] With regard to S-layers there are two notable observations:

1. So far, within the *Methanococcales* only pure protein S-layers have been found as cell wall components (for review see refs. 32 and 33), and

2. Within the *Methanobacteriales* there are the sheathed species *Methanospirillum hungatei* and *Methanosaeta* (formerly *Methanothrix*) sp. with a complex multilayered cell wall architecture (for review see refs. 34 and 35).

The hexagonally arrayed S-layer protein of *Methanococcus voltae* is preferentially released from the plasma membrane in HEPES buffer at 50 to 60°C.[36] Amino acid analyses showed a content of 25.3% acidic and 9.8% basic amino acids with no cysteine and a low amount of methionine. In accordance with these data a pI value of 4.0 was determined for the RS protein,[33] which appears characteristic for most S-layers proteins.[3]

The filamentous sheathed methanogen *Methanospirillum hungatei* synthesizes unusual extracellular crystalline macromolecular layers which have been investigated in great detail by Beveridge and coworkers in the past decade. Different components are involved in the assembly of these multilayered structures. The bacteria are encased in a filamentous sheath of unusual stability.[37] Neither an alkali dissolution technique nor a combined treatment using sodium dodecyl sulfate (SDS) and β-mercaptoethanol proved very effective to produce distinct solubilized sheath components.[38] However, by SDS/β-mercaptoethanol/2(N-cyclohexyl amino)ethanesulfonic acid or SDS/β-mercaptoethanol/EDTA dissolution procedures 42 and 74% of the mass of the sheath, respectively, were solubilized[39] and selected into a group of polypeptides of 10 to 40 kDa. Treatment of the sheath with 90% phenol resulted in the solubilization of a novel group of phenol-soluble polypeptides which accounted for ~19% of the mass of the sheath.[19] From these observations it was concluded that the sheath is a trilamellar structure in which phenol-soluble polypeptides are sandwiched between phenol-insoluble (SDS/β-mercaptoethanol/EDTA-soluble) polypeptides. Part of the tripartite envelope structure is the cell wall just on the inner face of the sheath which proved to be a typical S-layer. Characterization of this layer was very complicated due to its inherent instability after isolation.[20] Two major non-glycosylated polypeptides with molecular masses of 114 and 110 kDa, respectively, were extracted by an alkaline buffer treatment. Between the individual cells multilayered spacer plugs are located. They could be extracted by a novel freeze/crush/thaw method and dissociated in phosphate buffer (pH 7.4) at 56°C. So far, only an ultrastructural characterization has been performed.[40]

Rather similar sheath structures are present in *Methanosaeta* (formerly *Methanothrix*) *concilii* but have not yet been investigated in greater detail.[34] It appears that the functions of the particular envelope components differ in both organisms since their ways of substrate utilization and motility are different.[40]

A quite complex envelope structure was also described for *Methanosarcina* spp.[41] In this organism an ~100 nm thick methanochondroitin layer, composed primarily of a glucuronic acid and galactosamine containing heteropolysaccharide can be found on top of an ~10 nm thick S-layer, which is associated with the cytoplasmic membrane. After transition of aggregated cells to dis-aggregated single cells by high NaCl concentrations (> 0.4 M) in the growth medium only the S-layer remained intact but not the methanochondroitin layer. Polyvalent cations such as Mg^{2+} or spermine are required to maintain the integrity of the protein sub-units that compose the S-layer.[41]

Thermococcus stetteri[42] and *Thermococcus litoralis*[43] have been described to possess protein envelopes outside the plasma membrane. Recently, in a more detailed reinvestigation of the S-layer of *T. stetteri*[13] glycosylation was observed. Presumably the same is true for the closely related organism *Pyrococcus abyssi*.[44] It was suggested[29] that S-layers from organisms of the orders *Thermoproteales* and *Sulfolobales* are glycosylated. Since *Hyperthermus butylicus*[45] is part of a novel genus comprising the orders *Thermoproteales* and *Sulfolobales*[46] its S-layer protein is probably glycosylated, too. The same can be assumed for several other organisms of these orders including *Pyrobaculum islandicum*,[47] *P. organotrophum*,[12,47] *P. aerophilum*,[48] *Sulfolobus shibatae*,[49] *S. metallicus*,[50] *Acidianus brierleyi* (formerly *Sulfolobus brierleyi*),[51] *Metallosphaera sedula*,[52] and *Stygiolobus azoricus*.[53] The S-layers of either organism have been characterized only by electron microscopy but no details about molecular mass or chemical composition were presented. Definitely no carbohydrates have been found in the S-layer of *Sulfosphaerellus thermoacidophilum.*[54] The infrared spectrum (amide region) of the S-layer showed the occurrence of β-sheet structure.

3.2.2. EUBACTERIA

In the past few years the vast majority of reports on the chemical characterization of non-glycosylated eubacterial S-layers concerned about 10 different bacterial species. All of them are pathogens either for humans or animals and the fact that their S-layers are virulence factors caused them to be a subject of thorough investigations (see also chapters 5 and 7). For example, the involvement of the S-layer in the virulence of *Aeromonas salmonicida* has been published some 15 years ago.[55-57] The natural spread of aeromonads and isolation from different sources on the one hand and the enormous impact of modern analytical techniques, molecular biology and genetics on the other hand have tremendously extended our knowledge on chemical composition, amino acid sequence, secondary structure, functional domains, and strain-to-strain antigenic variations among S-layers of these organisms.

One of the best investigated systems is the S-layer of *Aeromonas salmonicida* composed of the so-called A-protein. In an earlier review Kay and Trust[58] have summarized their data on chemistry, three-dimensional structure, binding to LPS, sequence and function of the A-protein. The S-layer proteins of *Aeromonas salmonicida* and *Aeromonas hydrophila* are similar with respect to morphology and appear to be important virulence factors.[17,56,59] Various strains of *A. hydrophila* isolated from diseased and healthy fish were compared with regard to S-layers, serum sensitivity and pathogenicity in fish (for detailed description of the pathogenicity of *A. salmonicida* and *A. hydrophila* see chapter 5). Certain strains among them were shown to be serum resistant irrespective of whether they have S-layers or not. These observations suggest that S-layers do not always associate with protecting S-layer-carrying organisms from bactericidal activities.[60] Along this line an in vivo study in *A. salmonicida* shows that resistance to bacteriolysis and phagocytosis was associated with a newly acquired capsular layer.

Since this capsule shielded the A-layer the role of the A-layer in the overall pathogenesis might be less important than originally thought.[61]

A detailed biochemical analysis of the structural domains of the A-protein of *A. salmonicida* with 481 amino acid residues and a molecular weight of 50,800[62] shows that two major domains were present. Trypsin cleavage produced an N-terminal peptide of apparent M_r 39,400 which was totally refractile to trypsin, and a 16.7 kDa C-terminal peptide with intermediate resistance to trypsin.[24] The availability of the trypsin-resistant fragment enabled the comparison of the secondary structure of the intact A-protein with that of its major structural domain. Circular dichroism studies in the presence and absence of 0.1% SDS have shown that the secondary structures of both polypeptides were altered. Western and dot immunoblotting, immunomicroscopy, enzyme-linked immunosorbent assay with monospecific polyclonal antiserum and monoclonal antibodies specific for epitopes exposed on the surface of native A-layer revealed that ~70% of the protein was inaccessible for proteases or displayed non-epitopic residues. The majority of these residues was in the N-terminal 301 residues of the A-protein whereas the C-terminal 180 residues contained most of the surface-accessible sequences.[24] Analysis of a number of independent mutants of *A. salmonicida*, unable to produce intact S-layer protein at normal rates, demonstrated that they either synthesize A-layer subunits at a reduced rate or produce a truncated subunit, or completely lose the ability to produce the subunit protein. Responsible for this altered synthesis behavior of A-layer protein are two different insertion sequence elements inserted into different sites in the A-layer gene (*vapA*) and its promoter.[63] Another mutant of *A. salmonicida* (A449-TM1, a Tn5 mutant) was unable to secrete the A-layer protein through the outer membrane.[64] Instead, by labeling with polyclonal antisera against the S-layer protein the accumulation of large quantities of A-protein in an enlarged periplasm, preferably at the cell poles, has been demonstrated. However, the export of other extracellular proteins was not impaired, indicating a specific secretion pathway for the A-protein in *A. salmonicida*.[64] The surface array-producing strains of *A. salmonicida*,[65] *A. hydrophila*, and *A. veronii*[66] have a conserved property in common which has also been found in S-layer-carrying *Caulobacter crescentus* strains,[26] namely a lipopolysaccharide with O polysaccharides of homogenous length. These specific O chains are required for retaining the S-layer on the cell surface.[67,68]

The S-layer protein of *Aeromonas hydrophila* is structurally very similarly organized as the A-layer of *A. salmonicida.*[23] It contains also two morphological domains of comparable molecular masses. However, no *A. hydrophila* or *A. sobria* strain reacted with poly- and monoclonal antibodies against the *A. salmonicida* A-protein.[24] A mutant of *A. hydrophila* was isolated which produced a truncated S-layer protein of M_r 38,650 which neither could self-assemble into a regular lattice nor be anchored to the cell surface. By circular dichroism measurements it was demonstrated that the C-terminus was responsible for conformational changes of the S-layer protein upon environmental effects.[23] The truncated 38.6 kDa protein could be translocated across both the inner and outer membrane of the cell. However, due to the truncation it was also accumulated in the periplasm and the cytoplasm. The export of wild-type protein was obviously far more efficient,[69] suggesting the importance of the C-terminal domain in the export process.[23] Recently it was observed that the S-layer proteins of several investigated *A. hydrophila* strains contain phosphotyrosine.[70] Transformation of the *ahsA* gene encoding the S-layer protein of *A. hydrophila* TF7 into *Escherichia coli* led to stable high-level expression of the 448 residue 45.4 kDa mature protein with a predicted pI value of 6.72. This was in contrast to the measured M_r of 52,000 and pI value of 4.6.[17] The differences can be explained by a post-translational phosphorylation of tyrosine residues which was also confirmed by in vivo cell labeling with [^{32}P]orthophosphate.[70] Beside glycosylation[30,71] this phosphorylation event is the only post-translational modification of S-layers reported so far. The identification of post-translational modification, together with the gene sequence, corrects previous estimates of the amino acid content of *A. hydrophila* S-layer proteins.[17,72] These studies have shown strain-to-strain differences in N-terminal sequences, peptide polymorphism after endoproteinase Glu-C mapping, as well as antigenic differences among S-layer proteins of *A. hydrophila* and *A.veronii* biotype sobria.[73] Comparison of N-terminal sequences of

A. salmonicida and *A. hydrophila* strains TF7 and AH-342 showed homologies of only 4% to 24% between the sequences.[72] A detailed discussion of molecular biology and genetics of the surface arrays of *A. salmonicida* and *A. hydrophila* is provided in chapter 4.

Variation of S-layer protein antigenicity is also associated with *Campylobacter fetus* strains. Blaser and coworkers observed that *C. fetus* cells produce different S-layer proteins (M_r 98,000, 127,000 and 149,000).[74] On single cells different antigenetically crossreactive S-layer proteins have been found.[25] These high-molecular-weight S-layer proteins are the immunodominant antigens of *C. fetus*.[75] Usually only one of the different proteins is expressed on the cell surface, which was demonstrated with other *Campylobacter* isolates, too.[76,77] The proteins which possess conserved and variable regions within the same primary structure framework are generated by homologous recombination between expressed and silent copies of the S-layer protein gene.[78] For example, serotype A cells of *C. fetus* possess eight homologs of *sap*A, which encodes a 97-kDa S-layer protein. The gene products of these homologs have a conserved N-terminus of 184 amino acids. The serotype B gene is similar in structure to the type A gene *sap*A and encodes a full-length 936-amino acid (97 kDa) S-layer protein. However, sequence analysis of *sap*B indicated that the conserved N-terminal region in *sap*A was absent but the remaining 751 amino acids encoded by *sap*B were identical to that of *sap*A in spite of the non-conserved nature of this region among the *sap*A homologs.[79] In the previous studies an in vitro shift in S-layer protein expression was demonstrated.[78] Recently the same phenomenon has been observed in vivo in a natural host during infection. The in vivo antigenic variation is accompanied by genomic rearrangement and possible amplification of the *sap*A homologs.[80] A more comprehensive explanation of the genomic arrangements of the *sap*A homologs and the *sap*B gene is provided in chapter 4.

Rickettsia prowazekii and *Rickettsia typhi*, the etiologic agents of epidemic and endemic typhus, possess S-layer proteins with rather high molecular weights.[81] The *spa*P gene encoding the S-layer protein of *R. prowazekii* has recently been cloned, sequenced, and expressed in *E. coli*.[83] It encodes the surface array protein (SPA) of molecular weight 169,000. However, the surface forms of the

S-layers of both rickettsiae appear as 120 kDa proteins.[84] The gene for such a 120 kDa S-layer protein was recently characterized in *R. rickettsii*.[85] In order to study their immunochemistry purified S-layers of *R. prowazekii* and *R. typhi* were fragmented with cyanogen bromide (CNBr). The fragments were separated by SDS-PAGE, recovered after blotting, and the origins of the major fragments were determined by N-terminal sequencing.[84] The cleavage patterns and protein sequences of the two investigated proteins differ significantly. The obtained fragments were also analyzed by monoclonal antibodies of various selectivity and could be classified into eight different types. In each protein modified amino acids were detected (e.g. γ-N-methylasparagine and others). No fragments corresponding to the C-termini of both SPAs were found. In conclusion, this suggests that the C-terminal region either is not synthesized or the 200 to 300 C-terminal amino acids encoded by the *spaP* gene are removed during S-layer translocation to the cell surface.[84]

Clostridium difficile, a human pathogen causing pseudomembraneous colitis, is usually covered by two square S-layer lattices.[86] For the constituting proteins of *C. difficile* strain GAI 0714 respective molecular weights of 32,000 and 45,000 were found.[11,86-88] In an HPLC gel filtration experiment in the absence of SDS respective molecular weights of 61,000 and 99,000 have been determined, suggesting dimeric forms of the two proteins under non-denaturing conditions. The 32 kDa protein exhibits multiple isoelectric forms (pI values 3.7 to 3.9), whereas the 45 kDa protein reveals a single form (pI value 3.3).[11] N-terminal sequencing of the first 10 amino acids showed that only residue no. 4 (glutamic acid) and residue no. 9 (alanine) are identical in both S-layers. From comparison of their hydrophobicities it can be concluded that the 45 kDa protein is more hydrophobic that the other one. This is also reflected in the water solubility of either protein.[87] Self-assembly into regular arrays was dependent on the presence of divalent cations such as Ca^{2+} or Zn^{2+} but not on Ba^{2+} or Mg^{2+}. For self-assembly an equal mixture of both proteins was necessary. From inhibition studies with Fab fragments an intrinsic role of S-layers or the constituting proteins in in vivo bacterial adhesion was concluded.[87] Quite comparable results have been obtained from the analysis of the two S-layer proteins of *C. difficile* strain GAI 1152.[89] There is a shift in the molecular mass of the proteins from 32 and 45 kDa, respectively, to 38 and 42 kDa, respectively. The 38 kDa protein exhibits a single isoelectric form (pI value 4.0) but multiple forms are found with the 42 kDa protein (pI values 5.5 to 6.3). Peptide mapping results in rather different maps, indicating marked differences in the primary sequence of both S-layer constituents.[89]

The S-layer protein of *Clostridium tetani* strain AO 174, a non-toxigenic derivative of strain Harvard A 47 has been isolated by extraction with 4M urea and purified by a series of chromatographic steps.[90] By SDS-PAGE the molecular weight was determined to be 140,000 and amino acid analysis showed the typical composition of an S-layer protein[3] although the proline content was exceptionally low. Purified preparations showed multiple isoelectric forms ranging from pH 4.0 to 4.5.[90]

Clostridium botulinum type E Saroma was shown to possess an S-layer assembled of different subunits with molecular weights in the range of 10,000 to 150,000.[91] By immunoblot analysis using an antiserum against whole cells it was demonstrated that the 60 and 90 kDa subunits of the S-layer were the major somatic antigens of the organism. The amino acid compositions of the major subunits were typical for S-layer proteins.[91]

The *sap* gene encoding the S-layer protein of *Bacillus anthracis* 9131, the etiological agent of anthrax, was cloned in *E. coli* and the complete sequence of the structural gene was determined.[92] The calculated molecular mass is 83.7 kDa whereas a molecular mass of 94 kDa was determined by SDS-PAGE. The Sap protein possesses many charged residues, is weakly acidic and the content of sulfur-containing amino acids is rather low (0.9% methionine). This strain of *B. anthracis* is a derivative of the Sterne strain.[92] Another derivative is the avirulent *B. anthracis* strain Delta Sterne-1.[93] Its protein extractable antigen 1 (EA1) which is supposed to be the S-layer protein revealed only marginal differences in the molecular weight to the previously reported data.[92] The molecular mass of EA1 from strain Delta Sterne-1 but also from the attenuated *B. anthracis* Sterne is 95 kDa.[93] Under non-denaturing conditions gel permeation chromatography on a Superose 6 column yields a single peak at ~400 kDa indicating a monodisperse product in the form of either a tetramer or a dimer of dimers. Amino acid analysis showed that 47% of the total amino acids were nonpolar residues. The content of hydroxyl and basic amino acids was in the range of 15%, each. Surprisingly, N-terminal sequencing revealed an identical match with residues no. 2 to 8 of the S-layer protein of *Bacillus thuringiensis*.[94] In the absence of protease inhibitors during isolation of EA1 a specific proteolytic processing to an 80 kDa form occurred which still immunoreacted with polyclonal anti-EA1 antibodies.[93]

Among subgingival bacteria *Campylobacter rectus* (formerly *Wolinella recta*) is known for its relative abundance in periodontal lesion sites.[76,95] The acidic extracts of 23 isolates and strain ATCC 33238 contained mainly S-layer proteins consisting of two subunits.[95] Their molecular weights were 130 and 150 kDa, respectively. One isolate possessed a single S-layer of M_r 160,000. The antigenic properties of the 150 and 160 kDa proteins were investigated by immunodiffusion. The results indicated the existence of common antigenic determinants as well as strain (or group) specific antigens among the S-layer proteins. A variation in the isoelectric points between pI values of 5.8 to 7.2 has been observed. This pI variation suggests a different degree of deamidation and/or a possible alteration in surface exposed amino acids in the different S-layers.[95]

In the past five years a second group of eubacteria with nonglycosylated S-layer proteins has been investigated in greater detail because of their biotechnological application potential. It includes different lactic acid bacteria (e.g. *Lactobacillus helveticus*,[96] *L. acidophilus*,[16,97] *L. crispatus*,[98] propionibacteria,[99] *Bacillus stearo thermophilus*,[100] and *Caulobacter* strains).[101,102]

The first S-layer among lactic acid bacteria was observed on *Lactobacillus fermenti*.[103] Another S-layer-carrying organism is *Lactobacillus helveticus* which is particularly important in the dairy industry for cheese production. Lortal and coworkers have characterized the S-layer of *L. helveticus* ATCC 12046 in great detail.[96,104] The S-layer is a non-glycosylated 52 kDa protein with a content of 44% hydrophobic amino acids. Extraction of the S-layer protein was performed by 5M lithium chloride since this treatment leads only to limited loss of viability whereas extraction by 5M guanidine hydrochloride causes dramatic losses in cell viability.[96] A molecular mass of 43,533 has been determined by mass spectrometry.[105] The chemical compositions of both the intact cells and the isolated S-layers were further investigated by X-ray photoelectron spectroscopy.[106] As a result, the extracted protein was shown to be practically pure protein with only very small amounts of glycosidic residues.

Among 10 investigated *Lactobacillus acidophilus* strains six were found to be covered with an S-layer.[97] The apparent molecular weights varied from 41,000 to 49,000 as estimated by SDS-PAGE. Chemical cleavage by N-chlorosuccinimide and proteolysis with *Staphylococcus aureus* V8 protease resulted in markedly different peptide patterns, showing the inherent heterogeneity in these S-layers.[97] Pouwels and coworkers have cloned and sequenced the S-layer protein of *L. acidophilus* ATCC 4356.[16] The *slpA* gene encodes a protein of 444 amino acids including a 24 amino acid signal sequence. The molecular weight of the mature protein, determined by SDS-PAGE, is 43,000.[16] A quite similar S-layer gene was isolated from *L. brevis*.[15] It encodes a mature S-layer protein of 435 amino acids and a calculated M_r of 45,000. This is very close to the M_r of 46,000 determined by SDS-PAGE. Interestingly, for both proteins rather basic pI values of 9.4 for *L. acidophilus* and 9.88 for *L. brevis* were calculated.[15,16] *Lactobacillus acidophilus* and *L. crispastus* strains were assessed for the adherence of S-layers to proteins of the mammalian extracellular matrix.[98] Both strains possess S-layers with an M_r of 43,000 which were extracted with 2M guanidine hydrochloride. N-terminal sequencing of the *L. crispastus* S-layer protein resulted in the sequence DAVSSANNSNLGNV. This N-terminal sequence was compared with 20 S-layer protein sequences in the EMBL 35 and SWISS-PROT 26 data banks but exhibited neither homology to *L. acidophilus* ATCC 4356 S-layer protein[16] nor to any other S-layer sequence. Only the *L. crispatus* JCM 5810 S-layer adhered effectively to immobilized type IV and type I collagens, laminin, and with lower affinity to type V collagen and fibronectin. These observations indicate that *L. crispatus* obviously expresses a novel S-layer protein which exhibits not only structural heterogeneity in comparison to other S-layer proteins but also functional heterogeneity due to its collagen-binding properties.[98]

Among 70 investigated dairy propionibacteria only two S-layer-carrying organisms have been found.[99] *Propionibacterium freudenreichii* CNRZ 722 and *P. jensenii* CNRZ 87 possess oblique S-layers with molecular weights of 58.8 and 67.3 kDa, respectively, as was demonstrated by SDS-PAGE. Mass spectrometry of the S-layer of strain CNRZ 722 revealed a molecular mass of 56,533 Da. Amino acid compositions and N-terminal sequences of both proteins were different but the content of hydrophobic amino acids was about the same (52%).[99]

The chemical compositions of three different *Bacillus stearothermophilus* strains, two of them possessed non-glycosylated S-layer proteins and one S-layer was glycosylated, were analyzed during growth in continuous culture on complex medium under oxygen-limited conditions.[100] Briefly, when oxygen limitation during culture was relieved, physiological conditions changed dramatically and the different wild-type S-layer proteins were replaced in all cases by a new common type of oblique, non-glycosylated S-layer protein of M_r 97,000. A detailed analysis of the concomitant changes in the S-layers is provided in chapter 5.

Marine and freshwater caulobacters, particularly strains of *Caulobacter crescentus*, were the main research subject of Smit and coworkers for years. Chemical composition,[107] ultrastructure,[108] and genetics[109] have been described in great detail. In these previous studies a connection between shedding of RsaA protein and an S-layer-associated oligosaccharide has been made.[102] It was suggested that this homogenous-length lipopolysaccharides, termed SLPS, is required for S-layer surface attachment, possibly via calcium bridging[26] (see chapter 2). Its composition was tentatively assigned to be 4,6-dideoxy-4-aminohexose, 3,6-dideoxy-3-aminohexose, and glycerol in equal amounts.[26]

Taxonomical characterizations on several *Bacillus* species have established the occurrence of S-layer proteins in these organisms. *Bacillus brevis, B. migulanus, B. choshinensis, B. parabrevis,* and *B. galactophilus* are covered by hexagonal S-layers in the molecular weight range of 110,000 to 150,000.[110] In *B. aneurinolyticus,* however, two groups of organisms exist with either one S-layer protein

(115 kDa) or two cross-reactive S-layer proteins (105 and 115 kDa).[111] The S-layers found in the three newly described species *B. reuszeri*, *B. formosus*, and *B. borstelensis* were immunologically and genetically similar to those of *B. brevis* and related organisms.[112]

Characterization by SDS-PAGE of a peritrichously flagellated organism with a square S-layer lattice from a St. Lucia hot spring revealed the presence of a 83 kDa protein band which did not stain for carbohydrates. The S-layer of this organism resembles those of *Desulfotomaculum nigrificans* strains.[113,114]

3.3. GLYCOSYLATED S-LAYER PROTEINS

In this section the main focus of the review is directed towards glycosylation events on S-layer proteins and the determination of the structure of the S-layer glycan chains. The addition of covalently linked carbohydrate residues to a protein is a major post-translational modification.[115] Of particular interest is the fact that until recently it was believed that prokaryotes are not able to glycosylate their proteins (for review see ref. 116). In contrast to this general opinion Sumper and Wieland[117] published an excellent review on prokaryotic glycoproteins just in spring of 1995.

3.3.1. ARCHAEOBACTERIA

The first prokaryotic glycoprotein described was the S-layer glycoprotein of *Halobacterium salinarium*.[118] Sumper, Wieland and their coworkers have extended the biochemical and genetical studies on *Halobacterium halobium*. Three different types of N- and O-linked glycan chains have been identified on the halobacterial S-layer glycoprotein (Table 3.1). After characterization of the coding gene the biosynthesis of the glycans was investigated. It was demonstrated that instead of the well known bacterial lipid carrier undecaprenol a C_{60}-dolichol is involved in the biosynthetic pathway. Additionally, transient glycosylation of the halobacterial S-layer glycoprotein was observed. The results of these studies have been summarized in several reviews.[71,117,119]

In a parallel series of investigations the characterization of the S-layer glycoprotein of the moderate halophilic archaeobacterium *Haloferax volcanii* has been performed. After cloning and sequencing of the S-layer gene and determination of O-glycosylation sites,[120] which are quite similar to those of *H. halobium,*[71] dramatic differences were observed in the structure of the N-glycosidically linked saccharides.[121] Whereas in *H. halobium* sulfated hexuronic acid-containing oligosaccharides are attached via the novel N-glycosidic linkage unit asparaginyl-glucose to the S-layer polypeptide, in *H. volcanii* glucose-containing homopolysaccharides are bound via the same linkage type to the S-layer polypeptide.[121] The obvious differences between both S-layer glycoproteins are also reflected on DNA level. The proteins share only 41% similarity and *H. volcanii* DNA was not recognized by a *H. halobium*-specific probe.[21] One explanation for these differences is that the sulfated heteropolysaccharides of *H. halobium* add considerable amounts of negative charges to the protein which stabilizes the S-layer protein at high environmental salt concentrations (~4M NaCl). In *H. volcanii* which grows optimally at moderate salt concentrations (2.3M NaCl) such a stabilization effect by glycan chains is not required.[21] Similar stabilizing effects, based on the introduction of a large number of negative charges to a protein, have also been reported for a halophilic malate dehydrogenase.[122] Just recently novel glycoproteins and sulfated dihexosyl glycolipids were reported to occur in *H. volcanii.*[123] Photoaffinity labeling of a *H. volcanii* homogenate using 5-azido-[^{32}P]UDP-glucose tagged only one 45 kDa polypeptide. The fact that only one polypeptide band was photoaffinity-labeled indicated that no other transferase in *H. volcanii* directly utilizes UDP-glucose.[123]

In the cell wall of the triangular halophilic archaeobacterium *Haloarcula japonica* TR1 about 5% total carbohydrates were determined by the phenol-sulfuric acid method.[124,125] The sugars are components of the hexagonally arrayed S-layer glycoprotein but no further structural analyses have been performed.[124] Further, by N-terminal sequencing of the 180 kDa S-layer glycoprotein the 18 amino-terminal residues were determined.[125]

Among methanogenic bacteria most S-layer proteins appear to be glycosylated.[32] As already mentioned, most of the carbohydrate analyses are based on a periodic acid-Schiff staining reaction on SDS-polyacrylamide gels. Detailed genetical and structural analyses have been performed only on *Methanothermus fervidus.*[22,126] By nuclear magnetic resonance spectroscopy it was demonstrated that the heterosaccharide (Table 3.1) consists of 3-*O*-methyl-D-mannose, D-mannose and N-acetyl-D-galactosamine in a molar ratio of 2:3:1 and is linked via an N-glycosidic linkage from GalNAc to asparagine.[22] Some 3-*O*-methylmannose residues can be partly replaced by 3-*O*-methylglucose residues. For the hexasaccharide a molecular mass of 1,061.8 was calculated.

Biosynthesis studies on archaeobacterial S-layer glycans have been performed only on *Halobacterium halobium* and *Methanothermus fervidus* and are summarized in several reviews.[32,71,117,119,127] In the methanogens, for example, not only nucleotide-activated monosaccharides but also oligosaccharides are involved in the biosynthesis of the S-layer glycans.[32] Additionally, transient occurrence of specific sugar residues at specific biosynthetic stages was reported. These sugars are eventually cleaved off before final assembly of the mature S-layer glycoproteins. How and where these trimming reactions take place is not known yet.[32] In the methanogens both lipid carriers undecaprenol and dolichol have been found but only dolichol was involved in the biosynthesis of the S-layer glycoprotein.[127]

Among the methanogens glycosylated S-layers have been reported for a number of different species.[32] However, most characterizations are based only on carbohydrate staining reactions on SDS-polyacrylamide gels. For example, the hexagonally arranged 138 kDa S-layer glycoprotein of *Methanoculleus marisnigri* (formerly *Methanogenium marisnigri*) showed positive staining reaction either after Alcian blue or thymol-sulfuric acid treatment of gels.[128] Another organism is *Methanoplanus limicola*[129] whose hexagonal S-layer glycoprotein yielded a total content of neutral sugars of 24%. Upon treatment with trifluoromethanesulfonic acid the M_r, determined by SDS-PAGE, dropped significantly from 135,000 to 115,000 indicating a glycosylated S-layer protein.[129]

Among the Euryarchaeota[29] *Thermococcus stetteri* is the only archaeobacterium, except halophilic and methanogenic organisms, whose S-layer was investigated in more detail.[13] It is unique because it forms double layers. The two major proteins with apparent molecular weights of 80,000 and 210,000 were glycosylated as demonstrated by Alcian blue staining of SDS-polyacrylamide gels.[13]

Most of the S-layer-carrying archaeobacteria among the Crenarchaeota should possess glycosylated S-layer proteins.[29]

However, carbohydrates have only been reported experimentally in *Staphylothermus marinus*,[14] *Pyrodictium abyssi*,[130] and *Sulfurococcus mirabilis*.[131]

One of the most detailed analyses of the chemical composition of an S-layer-like glycoprotein (see chapter 2) has been carried out on *Staphylothermus marinus*.[14] The S-layer material was degraded by chemical and proteolytic methods into separate domains. The whole protein complex, termed tetrabrachion, revealed an overall carbohydrate content of 38%. After extraction with SDS complete S-layer "ghosts" were obtained which are essentially free of membrane components. The ghosts could be further broken down by different degradation methods into morphological subunits which tended to associate into rosette-shaped protein micelles via their membrane anchors. By treatment with organic solvents or with 0.5% SDS at elevated temperature those micelles could be further disintegrated; e.g. 70% formic acid yielded two major proteins with apparent molecular masses of 10 and 130 kDa. However, staining of the gels with colloidal Coomassie revealed additional bands at 85 kDa and at high apparent M_r. In total, five different protein domains were distinguished in the degraded tetrabrachion-protease complex.[14] Their specific carbohydrate contents varied between 18 and 53%. Due to limited amounts of degraded material no analyses of the carbohydrate structures could be performed.

After SDS-PAGE of cell envelopes of *Pyrodictium abyssi* one major periodate-Schiff positive protein band with an apparent molecular weight of 126,000 has been observed.[130] This band represents the glycosylated S-layer protein.

A more detailed characterization has been performed on the S-layer glycoprotein of *Sulfurococcus mirabilis*.[131] Amino acid analysis revealed increased amounts of aspartate, glutamate and the hydroxy-amino acids serine and threonine. On the other hand, the arginine content was rather low. By the phenol-sulfuric acid assay a carbohydrate content of the S-layer glycoprotein of ~10% was determined.[131]

3.3.2. EUBACTERIA

First evidence for the existence of glycosylated S-layers amongst eubacteria has been provided by Sleytr and Thorne.[132] They examined the hexagonal and square S-layer lattices of *Thermo-anaerobacter* (formerly *Clostridium*) *thermohydrosulfuricus* and *Clostridium thermosaccharolyticum* strains. Since then glycosylated eubacterial S-layer proteins have been reported almost exclusively on organisms belonging to the *Bacillaceae* family. However, among Gram-positive organisms a few exceptions such as *Sulfobacillus thermosulfidooxidans*[133] or *Corynebacterium glutamicum*[134] seem to exist. Whether S-layer glycoproteins are common structures in Gram-negative organisms (e.g. *Aquaspirillum sinuosum*[135]) remains to be established.

Aquaspirillum sinuosum cells are covered by two S-layer proteins.[135] The outer layer consists of a major 130 kDa protein and a 180 kDa minor component. These proteins are serologically related and show partial identity by peptide mapping. Periodic acid-Schiff staining of the 180 kDa band suggested that this could be a glycosylated form of the 130 kDa component.[135]

So far, complete structural analyses of glycosylated S-layer proteins from Gram-positive eubacteria have been exclusively described by our group.[1,30] All characterizations of glycan chains from the respective S-layer glycoproteins were carried out on glycopeptides derived by exhaustive pronase digestion. For purification of this material gel permeation chromatography, ion exchange chromatography, chromatofocusing and reversed phase HPLC were used (for example see ref. 27). The glycan structures of all investigated S-layer glycoproteins are summarized in Table 3.1. From these data it can be concluded that the glycans of most eubacterial S-layer glycoproteins consist of up to 50 repeating units. Their structures, however, differ considerably (see refs. 1 and 30, and Table 3.1). It is interesting to note that among the monosaccharide constituents of S-layer glycans sugars have been found which are typical representatives of O-antigens of lipopolysaccharides[136] such as quinovosamine[149] or D-rhamnose and N-acetyl-D-fucosamine.[31] Further support for the notion that S-layer glycoproteins of Gram-positive eubacteria and lipopolysaccharides of Gram-negative eubacteria are at least structurally related came from the recent observation of heptose residues as components of S-layer glycans. The repeating unit of a specific *B. thermoaerophilus* strain is composed of the disaccharide →4)-α-L-Rha*p*-(1→3)-β-

Table 3.1. Glycan structures of S-layer glycoproteins[1,30]

Eubacteria:
Bacillus stearothermophilus NRS 2004/3a[137,139]

[→2)-α-L-Rhap-(1→2)-α-L-Rhap-(1→3)-β-L-Rhap-(1→]n-50   Rhap-(1→N)-Asn

[→4)-β-ManpA2,3diNAc-(1→3)-α-GlcpNAc-(1→4)-β-ManpA2,3diNAc-(1→6)-α-Glcp-(1→]n-15

Bacillus thermoaerophilus L420-91[31] (Meier-Stauffer K, Busse H-J, Rainey FA et al. Int J Syst Bacteriol; submitted)
[→3)-α-D-Rhap-(1→3)-α-D-Rhap-(1→2)-α-D-Rhap-(1→2)-α-D-Rhap-(1→]n-15

```
                       2               2
                       ↑               ↑
                       1               1
              α-D-Fucp3NAc    α-D-Fucp3NAc
```

Bacillus thermoaerophilus DSM 10155 (formerly *Bacillus brevis* ATCC 12990)[140]
(Meier-Stauffer K, Busse H-J, Rainey FA et al. Int J Syst Bacteriol; submitted)
[→4)-α-L-Rhap-(1→3)-β-D-glycero-D-manno-Hepp-(1→]n

Paenibacillus alvei CCM 2051 (formerly *Bacillus alvei*)[28,141]

```
                                  α-D-Glcp
                                     1
                                     ↓
                                     6
```
β-D-Galp-(1→4)-β-D-ManpNAc-(1→[3]-β-D-Galp-(1→4)-β-D-ManpNAc-(1→]n-20 3)-
α-L-Rhap-(1→3)-α-L-Rhap-(1→3)-α-L-Rhap-(1→3)-β-D-Galp-(1→O)-Tyr
```
                                     4
                                     ↑
                                     1
        GroA-(2→O)-PO2-(O→4)-β-D-ManpNAc
```

Thermoanaerobacter thermohydrosulfuricus L111-69 (formerly *Clostridium thermohydrosulfuricum*)[27,142]   ~4 copies
3-OMe-α-L-Rhap-(1→4)-α-D-Manp-(1→[3]-α-L-Rhap-(1→4)-α-D-Manp-(1→]n-27 3)-
α-L-Rhap-(1→3)-α-L-Rhap-(1→3)-α-L-Rhap-(1→3)-β-D-Galp-(1→O)-Tyr

Thermoanaerobacter thermohydrosulfuricus L110-69 (formerly *Clostridium thermohydrosulfuricum* DSM 568)[27]   ~4 copies

3-OMe-α-L-Rhap-(1→4)-α-D-Manp-(1→[3)-α-L-Rhap-(1→4)-α-D-Manp-(1→]ₙ₋₂₇3)-

α-L-Rhap-(1→3)-α-L-Rhap-(1→3)-α-L-Rhap-(1→3)-β-D-Galp-(1→O)-Tyr

Thermoanaerobacter thermohydrosulfuricus S102-70 (formerly *Clostridium thermohydrosulfuricum*)[143,144]   ~4 copies

β-D-Galf-(1→3)-α-D-Galp-(1→2)-α-L-Rhap-(1→3)-α-D-Manp-(1→3)-α-L-Rhap-(1→3)-β-D-Glcp-(1→O)-Tyr

Thermoanaerobacter thermohydrosulfuricus L77-66 (formerly *Clostridium thermohydrosulfuricum* DSM 569)[145]

[→3)-α-D-GalpNAc-(1→3)-α-D-GalpNAc-(1→]ₙ₋₂₅   O-glycosidic bond via Tyr ?

4
↑
1

α-D-GlcpNAc-(1→2)-β-D-Manp

Thermoanaerobacter thermohydrosulfuricus L92-71 (formerly *Clostridium thermohydrosulfuricum*) (Christian R, Messner P; unpublished)

[→3)-α-D-GalpNAc-(1→3)-α-D-GalpNAc-(1→]ₙ₋₂₅   O-glycosidic bond via Tyr ?

4
↑
1

α-D-GlcpNAc-(1→2)-β-D-Manp

Thermoanaerobacter kivui DSM 2030 (formerly *Acetogenium kivui*)[146,147]

Heterosaccharide→Tyr   4 copies

Clostridium thermosaccharolyticum D120-70[148]

[→3)-β-D-Manp-(1→4)-α-L-Rhap-(1→3)-α-D-Glcp-(1→4)-α-L-Rhap-(1→]ₙ

2
↑
1

α-D-Galp

β-D-Glcp
         4
[→4)-β-D-GlcpNAc-(1→3)-β-D-ManpNAc-(1→]ₙ   ↑
         1

(α-D-Galp)₀.₅

Table continues on next page

Table 3.1. (Continued) Glycan structures of S-layer glycoproteins[1,30]

Clostridium thermosaccharolyticum E207-71[149]

$[\rightarrow 4]$-β-D-Galp-$(1\rightarrow 4)$-β-D-Glcp-$(1\rightarrow 4)$-α-D-Manp-$(1\rightarrow)]_{n\sim17}$ O-glycosidic bond via Tyr ?

$\overset{\displaystyle 3}{\underset{\displaystyle \uparrow}{}}$
$\overset{\displaystyle 1}{}$

β-D-Quip3NAc-$(1\rightarrow 6)$-β-D-Galf-$(1\rightarrow 4)$-α-L-Rhap
$[\rightarrow 4]$-β-D-GlcpNAc-$(1\rightarrow 3)$-β-D-ManpNAc-$(1\rightarrow)]_n$ (Schäffer C, Altman E, Messner P; unpublished)

$\overset{\displaystyle 4}{\underset{\displaystyle \uparrow}{}}$
$\overset{\displaystyle 1}{}$

$(\text{α-D-Rib}f)_{0.5}$

Clostridium symbiosum HB25[150]

$[\rightarrow 6]$-α-D-ManpNAc-$(1\rightarrow 4)$-β-D-GalpNAc-$(1\rightarrow 3)$-α-D-BacpNAc-$(1\rightarrow 4)$-α-D-GalpNAc-$(1\rightarrow O)$-PO$_2$-$(O\rightarrow)]_{n\sim15}$

Lactobacillus buchneri 41021/251[151]

α-D-Glcp-$(1\rightarrow[6)$-α-D-Glcp-$(1\rightarrow)]_{n\approx 6}$-α-D-Glcp-$(1\rightarrow O)$-Ser

Archaeobacteria:
Halobacterium halobium R$_1$M$_1$[21,71]

OSO$_3^-$	OSO$_3^-$	Ala-NH$_2$

$[\rightarrow 4]$-GlcNAc-$(1\rightarrow 4)$-GalA-$(1\rightarrow 3)$-GalNAc-$(1]_{n\sim 10\text{-}15}\rightarrow N)$-Asn   1 copy

$\overset{\displaystyle 6}{\underset{\displaystyle \uparrow}{}}\quad\overset{\displaystyle 3}{\underset{\displaystyle \uparrow}{}}$
$\overset{\displaystyle 1}{}\quad\overset{\displaystyle 1}{}$

3-OMe-GalA   Galf

Ala
|
Ser
|

GlcA-(1→4)-GlcA-(1→4)-GlcA-(1→4)-β-D-Glc-(1→N)-Asn
  | | | |
OSO_3^- OSO_3^- OSO_3^- X
                             Thr/Ser   ~10 copies

α-D-Glc-(1→2)-Gal-(1→O)-Thr   ~20 copies

Haloferax volcanii DS2[21,120,121]

β-D-Glc-(1→[4]-β-D-Glc-(1→]ₙ₋₈₋4)-β-D-Glc-(1→N)-Asn   ~4-7 copies

Glc-(1→2)-Gal-(1→O)-Thr   ~5-10 copies

Methanothermus fervidus V24S[22]

3-OMe-α-D-Manp-(1→6)-3-OMe-α-D-Manp-(1→[2]-α-D-Manp-(1→]ₘ₋₃4)-D-GalNAc-(1→N)-Asn   ~10 copies

Methanosaeta soehngenii FE (formerly *Methanothrix soehngenii*)[152]

Oligosaccharide-Rha-(1→N)-Asn

Abbreviations: Glcp, glucopyranose; Galf, galactofuranose; Man, mannose; Rha, rhamnose; Hep, heptose; Rib, ribose; GlcNAc, N-acetyl glucosamine; GalNAc, N-acetylgalactosamine; ManNAc, N-acetylmannosamine; Fuc3NAc, 3-N-acetylfucosamine (3-acetamido-3,6-dideoxyglucose); BacNAc, N-acetylbacillosamine (2-acetamido-4-amino-2,4,6-trideoxyglucose); Qui3NAc, 3-N-acetylquino vosamine (3-acetamido-...); ManA, mannuronic acid; GalA, galacturonic acid; GlcA, glucuronic acid; OMe, O-methyl; PO_4, phosphate; OSO_3^-, sulfate; Asn, asparagine; Tyr, tyrosine; Ser, serine; Thr, threonine; Ala, alanine.

D-*glycero*-D-*manno*- Hep*p*-(1→.[140] In contrast to the common
L-*glycero*-D-*manno*-heptose this heptose residue is in the D-*glycero*-
configuration which has been frequently observed in O-antigens
of LPS.[136]

Characterization of the carbohydrate-protein linkage regions of
different thermophilic and mesophilic *Bacillus* and *Thermoanaerobacter* (formerly *Clostridium*) strains also lead to the detection
of novel linkage regions. The first observation of a tyrosine-
linked glycan chain has been reported in *Thermoanaerobacter
thermohydrosulfuricus* S102-70.[143] In contrast to the common
polysaccharide chains (Table 3.1) the S-layer glycans of this
strain consist of hexasaccharides which are attached by an alkali-
stable O-glycosidic linkage between β-D-glucose and tyrosine to
the S-layer polypeptide. Calculation of the number of glycosylation
sites indicated that four to five different sites are present on
the S-layer protein protomer.[143] The number of glycosylation
sites was substantiated by results of glycan analyses from
T. thermohydrosulfuricus L111-69 and L110-69.[27] After proteolytic digestion of the intact S-layer glycoprotein four different
glycopeptides have been isolated which showed different amino
acid compositions of their peptide portions but their linkage amino
acid tyrosine was always present. Based on the different hydrophobicity properties the glycopeptides were separated by reversed
phase HPLC.[27] The glycan chains share identical constituents
but the number of disaccharide repeats of the structure →3)-α-L-
Rha*p*-(1→4)-α-D-Man*p*-(1→ varied between 23 and 33 with a
maximum at 28 repeats. At the non-reducing end the chain is
capped by a repeating unit with a modified rhamnose residue
(3-*O*-methylrhamnose). The whole polysaccharide chain is linked
to the S-layer protein via a core consisting of three α1,3-linked
rhamnoses and one β-D-galactose residue which thereupon is bound
to tyrosine residues of the S-layer protein.[27] This β-D-galactose-
tyrosine linkage was not described before. However, tyrosine-linked
glycan chains have also been found in the S-layer glycoprotein
of *Thermoanaerobacter* (formerly *Acetogenium*) *kivui* but glycan
structure and linkage sugars of this glycoprotein are not known
yet.[146,147] In this organism four glycosylation sites per protomer were
deduced from sequencing experiments.

Of particular interest was the observation that in *Paenibacillus* (formerly *Bacillus*) *alvei* CCM 2051 the S-layer glycan is linked via a similar core structure as in *T. thermohydrosulfuricus* L111-69[28] to the polypeptide, although the repeating units of both organisms possess completely different structures. A general scheme about the structural organization of S-layer glycoproteins can be drawn from this observation which resembles the assembly of O-antigens of LPS. Strain-specific oligosaccharide chains are attached via identical cores to the S-layer protein (Fig. 3.1). However, it is important to mention that not all eubacterial S-layer glycans are linked via tyrosine residues to the respective S-layer proteins. In *Bacillus thermoaerophilus* L420-91,[31] for example, the glycans are probably linked by O-glycosidic linkages to threonine residues of the S-layer polypeptide. Detailed analyses including the characterization of the core structure are currently being performed (Neuninger C, Messner P; unpublished observation).

In other Gram-positive eubacteria only less detailed characterizations of glycosylated S-layer proteins have been performed. For example, chemical characterization of a peritrichously flagellated organism with a hexagonal S-layer lattice from St. Lucia hot springs by SDS-PAGE and subsequent staining by Alcian blue or thymolsulfuric acid methods has revealed the presence of an S-layer glycoprotein with an apparent M_r of 200,000.[113] Physiological tests suggested that this organism belongs to the genus *Clostridium*.

The S-layer of *Sulfobacillus thermosulfidooxidans* is a glycoprotein containing ~10% carbohydrates.[133] Mannose was the main component but there were also decreasing amounts of glucosamine, glucose, xylose and galactose. The amino acid composition was found to be typical for an S-layer protein with a large proportion of hydrophobic residues, 21.5% acidic and 8.2% basic amino acids.[3,133]

Characterization of the *cspB* gene encoding PS2, the S-layer protein of *Corynebacterium glutamicum*, revealed a molecular weight of ~63,000 for the mature protein.[134] However, the 510 amino acid-polypeptide yielded a calculated molecular mass of 55,426 Da. From the nucleotide sequence of the *cspB* gene a content of 45.2% hydrophobic, 17.7% acidic, and of 8.3% basic amino acids was deduced resulting in a pI value of 4.45. In the sequence seven potential glycosylation sites were present but the definite occurrence

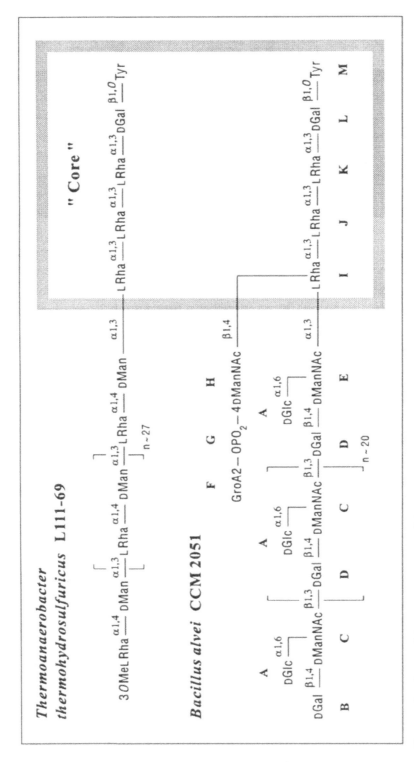

Fig. 3.1. Similarity of the core structures of the S-layer glycoproteins of Thermoanaerobacter thermohydrosulfuricus L111-69 and Paenibacillus (formerly Bacillus) alvei CCM 2051. (Reproduced from Messner et al. J Bacteriol 1995; 177:2188-93 with permission from the American Society for Microbiology.)

of glycan chains remains to be established which could explain the shift differences between calculated and observed M_r.[134]

Labeling of the glycan chains of *T. thermohydrosulfuricus* L111-69 after chemical modification of the hydroxyl groups into carboxyl groups, activation, and binding of ferritin has demonstrated that the carbohydrate chains protrude from the cell surface some 30 to 40 nm.[153] Charge interactions, as were demonstrated in the S-layer glycoprotein of *Clostridium symbiosum* HB25,[150] may

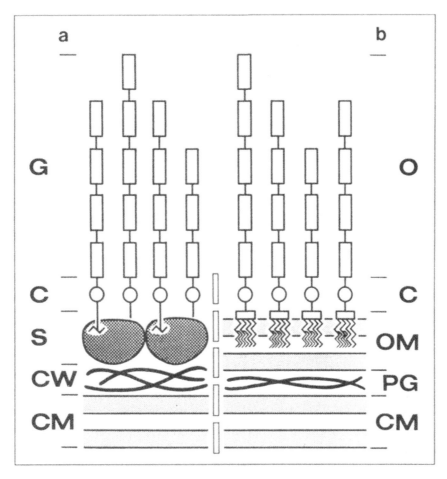

Fig. 3.2. Schematic presentation of the cell surface of (a) Gram-positive eubacteria carrying S-layer glycoproteins and (b) Gram-negative eubacteria with lipopolysaccharides. In both examples the glycan chains form the outermost layer of the cell. CM, cytoplasmic membrane; CW, cell wall; S, S-layer protein; C, core sugars; G, S-layer glycoprotein glycan chain assembled of repeating units; PG, peptidoglycan; OM, outer membrane; O, LPS O-antigen assembled of repeating units.

participate in the interaction to neighboring cells as well as to the surrounding environment. For a prokaryotic organism the S-layer glycans and the LPS O-antigens do create a comparable hydrophilic environment on the cell surface (Fig. 3.2). Therefore, investigations on the significance of glycosylation of S-layer proteins should also incorporate the perspective of functional similarities between these structures. Whether the general structural similarities between S-layer glycoproteins of Gram-positive eubacteria and the O-antigens of LPS of Gram-negative eubacteria are also reflected in common biosynthetic pathways for both types of macromolecules is under investigation.

The characterization of the biosynthesis of S-layer glycan chains in eubacteria has been initiated on *Paenibacillus* (formerly *Bacillus*) *alvei* CCM 2051.[154] Isolation and characterization of activated intermediates of the biosynthetic pathway from the cytoplasm indicated that nucleotide-activated oligosaccharides were formed from nucleotide-bound monosaccharides. This is a common observation among different S-layer glycoproteins of archaeobacteria and eubacteria.[32] However, there is a remarkable difference with regard to other glycoconjugates where oligosaccharides are only found in the lipid-bound state. As carrier lipid C_{55}-dolichol, was identified instead of the common prokaryotic lipid carrier undecaprenol. In both activated precursor forms additional sugars have been found which were not present in the mature glycan chains of the S-layer glycoprotein of *P. alvei*.[154] Where this trimming reaction occurs is presently not known. Similar biosynthetic routes for the assembly of S-layer glycan chains have been suggested also for other eubacteria such as *Clostridium thermosaccharolyticum* E207-7[149,155] but final data are not yet available. The only general conclusion which can be drawn presently is, that for the biosynthesis of eubacterial S-layer glycans nucleotide-activated oligosaccharides play an important role. Further, dolichol rather than undecaprenol is the common lipid carrier in eubacterial S-layer glycoprotein biosynthesis.[32,154]

3.4. CONCLUSIONS

In the past biochemical methods combined with data from electron microscopical investigations have provided basic knowledge on ultrastructure and chemical composition of eubacterial and archaeobacterial S-layer proteins (for review see refs. 1-4,6,107,156 and 157). Most S-layers consist of a single protein or glycoprotein species[1,3] but recent analyses have convincingly demonstrated that some of them contain more than one isoelectric protein form of the S-layer protein.[11,89,90] A rather complex S-layer-like structure, consisting of a number of different glycosylated polypeptides was recently described in *Staphylothermus marinus*.[14] In addition to S-layers built up of several subunits, modifications of the S-layer protein such as glycosylation[1,30,31] or phosphorylation[70] have been found. Despite the above mentioned variations of S-layer proteins no dramatic differences in the overall chemical composition of the protein portion of S-layers from different bacteria have been observed. In contrast, remarkable differences were determined for the molecular masses of the constituting subunits and the lattice types and lattice dimensions of S-layers (see ref. 1 and chapter 2).

By application of molecular biological and genetical methods to characterize S-layer proteins, it became clear very soon that, in general, the homology between S-layer proteins of different bacteria is not very high.[1,134,158] These findings confirm previous biochemical and ultrastructural observations about the diversity of S-layers and indicate specific functional roles for S-layers.

A complex modification of S-layer proteins is their glycosylation. S-layers were the first prokaryotic glycoproteins described.[118] Since then a number of eubacterial and archaeobacterial S-layer glycan structures have been analyzed (Table 3.1). Apart from the existence of S-layer glycoproteins an increasing number of other prokaryotic glycoproteins (non-S-layer glycoprotein) is now described in recent literature.[159] Of particular interest is that many S-layer glycan chains consist of repeating units which resemble in their structure O-antigens of lipopolysaccharides of Gram-negative eubacteria.[136]

Recent analyses of linkage regions of S-layer glycoproteins have shown the occurrence of both well known but also completely new linkage types. Due to the very limited knowledge about their structures in eubacteria and archaeobacteria generalizations are still rather difficult. One conclusion, however, is that in eubacteria apparently O-linked long-chain glycans do dominate (Table 3.1). Up to now there is only one example either of short heterosaccharide in *T. thermohydrosulfuricus* S102-70[143,144] or of an N-linked glycan chain in *B. stearothermophilus* NRS 2004/3a.[139] In contrast, in archaeobacteria N-glycosidically linked short heterosaccharides seem to be the preponderant glycan species (Table 3.1). Presently, no O-linked archaeobacterial S-layer glycans are known. An interesting observation has been made among S-layer glycoproteins of different *T. thermohydrosulfuricus* species. So far, all glycans investigated seem to be linked through the unusual O-glycosidic linkage via tyrosine to the S-layer polypeptide.[1,27] Although this linkage type seems to be strain-specific for *T. thermohydrosulfuricus* strains, a similar linkage region was found in *Paenibacillus alvei*[28] (Fig. 3.1).

Detailed analyses of the biosynthesis of S-layer glycoproteins have only been performed on the archaeobacteria *Halobacterium halobium*[71,117] and *Methanothermus fervidus*.[127] Nucleotide-linked oligosaccharides and new lipid carriers play an important role.[154] In *H. halobium* the oligosaccharides are completed and sulfated while still attached to dolichol on the cytosolic side of the cell membrane. Then, they are translocated in the lipid-bound state to the cell surface, possibly by a mechanism which involves transient methylation of the peripheral glucose residues. The S-layer protein can be translocated in the unglycosylated form through the cell membrane. Thus, despite the lack of relevant compartments such as the Golgi apparatus halobacterial glycoprotein biosynthesis in essence resembles the glycoprotein biosynthesis in eukaryotes.[117] No data are known yet about glycoprotein biosynthesis in eubacteria. An understanding of the biosynthetic pathway involved in S-layer glycoprotein biosynthesis first requires that a number of components have to be established. Once individual glycosidation steps have been characterized at the biochemical level, the next challenge will be to assemble the S-layer glycoproteins from component parts.

ACKNOWLEDGMENTS

I would like to thank Profs. U.B. Sleytr, F.M. Unger and P. Kosma for valuable discussions, Helene Hendling for the help with the preparation of the manuscript, and my coworkers for their support. This work was made possible by grants from the Austrian

Science Foundation, project S7201-MOB and the Austrian Federal Ministry of Science, Research and the Arts.

REFERENCES

1. Messner P, Sleytr UB. Crystalline bacterial cell-surface layers. In: Rose AH, ed. Advances in Microbial Physiology. Vol. 33. London: Academic Press, 1992:213-75.
2. Sleytr UB, Messner P, Pum D et al. Crystalline bacterial cell surface layers. Mol Microbiol 1993; 10:911-16.
3. Sleytr UB, Messner P. Crystalline surface layers on bacteria. Annu Rev Microbiol 1983; 37:311-39.
4. Kandler O, König H. Cell envelopes of archaebacteria. In: Woese CR, Wolfe RS, eds. The Bacteria. Vol VIII. Archaebacteria. New York: Academic Press, 1985:413-57.
5. Sleytr UB, Messner P, Pum D et al, eds. Crystalline Bacterial Cell Surface Layers. Berlin: Springer, 1988.
6. Beveridge TJ. Bacterial S-layers. Curr Opin Struct Biol 1994; 4:204-12.
7. Kist ML, Murray RGE. Components of the regular surface array of *Aquaspirillum serpens* MW5 and their assembly in vitro. J Bacteriol 1984; 157:599-606.
8. Austin JW, Murray RGE. Isolation and in vitro assembly of the components of the outer S-layer of *Lampropedia hyalina*. J Bacteriol 1990; 172:3681-89.
9. Yamada H, Tsukagoshi N, Udaka S. Morphological alterations of cell wall concomitant with protein release in a protein-producing bacterium, *Bacillus brevis* 47. J Bacteriol 1981; 148:322-32.
10. Sára M, Moser-Thier K, Kainz U et al. Characterization of S-layers from mesophilic bacillaceae and studies on their protective role towards muramidases. Arch Microbiol 1990; 153:209-14.
11. Takeoka A, Takumi K, Koga T et al. Purification and characterization of S layer proteins from *Clostridium difficile* GAI 0714. J Gen Microbiol 1991; 137:261-67.
12. Phipps BM, Huber R, Baumeister W. The cell envelope of the hyperthermophilic archaebacterium *Pyrobaculum organotrophum* consists of two regularly arrayed protein layers: three-dimensional structure of the outer layer. Mol Microbiol 1991; 5:253-65.
13. Gongadze GM, Kostyukova AS, Miroshnichenko ML et al. Regular proteinaceous layers of *Thermococcus stetteri* cell envelope. Curr Microbiol 1993; 27:5-9.
14. Peters J, Nitsch M, Kühlmorgen B et al. Tetrabrachion: a filamentous archaebacterial surface protein assembly of unusual structure and extreme stability. J Mol Biol 1995; 245:385-401.

15. Vidgrén G, Palva I, Pakkanen R et al. S-layer protein gene of *Lactobacillus brevis*: cloning by polymerase chain reaction and determination of the nucleotide sequence. J Bacteriol 1992; 174:7419-27.

16. Boot HJ, Kolen CPAM, van Noort JM et al. S-layer protein of *Lactobacillus acidophilus* ATCC 4356: purification, expression in *Escherichia coli*, and nucleotide sequence of the corresponding gene. J Bacteriol 1993; 175:6089-96.

17. Dooley JSG, McCubbin WD, Kay CM et al. Isolation and biochemical characterization of the S-layer protein from a pathogenic *Aeromonas hydrophila* strain. J Bacteriol 1988; 170:2631-38.

18. Dubreuil JD, Logan SM, Cubbage S et al. Structural and biochemical analyses of a surface array protein of *Campylobacter fetus*. J Bacteriol 1988; 170:4165-73.

19. Southam G, Beveridge TJ. Characterization of novel, phenolsoluble poly peptides which confer rigidity to the sheath of *Methanospirillum hungatei* GP1. J Bacteriol 1992; 174:935-46.

20. Firtel M, Southam G, Harauz G et al. Characterization of the cell wall of the sheathed methanogen *Methanospirillum hungatei* GP1 as an S layer. J Bacteriol 1993; 175:7550-60.

21. Sumper M. S-layer glycoproteins from moderately and extremely halophilic archaeobacteria. In: Beveridge TJ, Koval SF, eds. Advances in Bacterial Paracrystalline Surface Layers. New York: Plenum, 1993:109-117.

22. Kärcher U, Schröder H, Haslinger E et al. Primary structure of the heterosaccharide of the surface glycoprotein of *Methanothermus fervidus*. J Biol Chem 1993; 268:26821-26.

23. Thomas S, Austin JW, McCubbin WD et al. Roles of structural domains in the morphology and surface anchoring of the tetragonal paracrystalline array of *Aeromonas salmonicida*: biochemical characterization of the major structural domain. J Mol Biol 1992; 228:652-61.

24. Doig P, McCubbin WD, Kay CM et al. Distribution of surface-exposed and non-accessible amino acid sequences among the two major structural domains of the S-layer protein of *Aeromonas salmonicida*. J Mol Biol 1993; 233:753-65.

25. Fujimoto S, Takade A, Amako K et al. Correlation between molecular size of the surface array protein and morphology and antigenicity of the *Campylobacter fetus* S-layer. Infect Immun 1991; 59:2017-22.

26. Walker SG, Karunaratne DN, Ravenscroft N et al. Characterization of mutants of *Caulobacter crescentus* defective in surface attachment of the paracrystalline surface layer. J Bacteriol 1994; 176:6312-23.

27. Bock K, Schuster-Kolbe J, Altman E et al. Primary structure of the *O*-glycosidically linked glycan chain of the crystalline surface layer glycoprotein of *Thermoanaerobacter thermohydrosulfuricus* L111-69. Galactosyl tyrosine as a novel linkage unit. J Biol Chem 1994; 269:7137-44.

28. Messner P, Christian R, Neuninger C et al. Similarity of "core" structures in two different glycans of tyrosine-linked eubacterial S-layer glycoproteins. J Bacteriol 1995; 177:2188-93.

29. Kandler O. Cell wall biochemistry and three-domain concept of life. System Appl Microbiol 1994; 16:501-9.

30. Messner P, Sleytr UB. Bacterial surface layer glycoproteins. Glycobiology 1991; 1:545-51.

31. Kosma P, Neuninger C, Christian R et al. Glycan structure of the S-layer glycoprotein of *Bacillus* sp. L420-91. Glycoconjugate J 1995; 12:99-107.

32. König H, Hartmann E, Kärcher U. Pathways and principles of the biosynthesis of methanobacterial cell wall polymers. System Appl Microbiol 1994; 16:510-17.

33. Jarrell KF, Koval SF. Ultrastructure and biochemistry of *Methanococcus voltae*. CRC Crit Rev Microbiol 1989; 17:53-87.

34. Beveridge TJ, Patel GB, Harris BJ et al. The ultrastructure of *Methanothrix concilii*, a mesophilic aceticlastic methanogen. Can J Microbiol 1986; 32:703-10.

35. Beveridge TJ, Graham LL. Surface layers of bacteria. Microbiol Rev 1991; 55:684-705.

36. Koval SF, Jarrell KF. Ultrastructure and biochemistry of the cell wall of *Methanococcus voltae*. J Bacteriol 1987; 169:1298-306.

37. Beveridge TJ, Stewart M, Doyle RJ et al. Unusual stability of the *Methanospirillum hungatei* sheath. J Bacteriol 1985; 162:728-37.

38. Sprott GD, Beveridge TJ, Patel GB et al. Sheath disassembly in *Methano spirillum hungatei* strain GP1. Can J Microbiol 1986; 32:847-54.

39. Southam G, Beveridge TJ. Dissolution and immunochemical analysis of the sheath of the archaeobacterium *Methanospirillum hungatei* GP1. J Bacteriol 1991; 173:6213-22.

40. Firtel M, Southam G, Harauz G et al. The organization of the paracrystalline multilayered spacer-plugs of *Methanospirillum hungatei*. J Struct Biol 1994; 112:160-71.

41. Sowers KR, Boone JE, Gunsalus RP. Disaggregation of *Methanosarcina* spp. and growth as single cells at elevated osmolarity. Appl Environ Microbiol 1993; 59:3832-39.

42. Miroshnichenko ML, Bonch-Osmolovskaya EA, Neuner A et al. *Thermococcus stetteri* sp. nov., a new extremely thermophilic marine sulfur-metabolizing archaebacterium. System Appl Microbiol 1989; 12:257-62.

43. Neuner A, Jannasch HW, Belkin S et al. *Thermococcus litoralis* sp. nov.: a new species of extremely thermophilic marine archaebacteria. Arch Microbiol 1990; 153:205-7.

44. Erauso G, Reysenbach A-L, Godfroy A et al. *Pyrococcus abyssi* sp. nov., a new hyperthermophilic archaeon isolated from a deep-sea hydrothermal vent. Arch Microbiol 1993; 160:338-49.

45. Baumeister W, Santarius U, Volker S et al. The surface protein of *Hyperthermus butylicus*: three-dimensional structure and comparison with other archaebacterial surface proteins. System Appl Microbiol 1990; 13:105-11.

46. Zillig W, Holz I, Janekovic D et al. *Hyperthermus butylicus*, a hyperthermophilic sulfur-reducing archaebacterium that ferments peptides. J Bacteriol 1990; 172:3959-65.

47. Huber H, Kristjansson JK, Stetter KO. *Pyrobaculum* gen. nov., a new genus of neutrophilic, rod-shaped archaebacteria from continental solfataras growing optimally at 100°C. Arch Microbiol 1987; 149:95-101.

48. Völkl P, Huber R, Drobner E et al. *Pyrobaculum aerophilum* sp. nov., a novel nitrate-reducing hyperthermophilic archaeum. Appl Environ Microbiol 1993; 59:2918-26.

49. Grogan D, Palm P, Zillig W. Isolate B12, which harbours a virus-like element, represents a new species of the archaebacterial genus *Sulfolobus, Sulfolobus shibatae*, sp. nov. Arch Microbiol 1990; 154:594-99.

50. Huber G, Stetter KO. *Sulfolobus metallicus*, sp. nov., a novel strictly chemolithoautotrophic thermophilic archaeal species of metal-mobilizers. System Appl Microbiol 1991; 14:372-78.

51. Baumeister W, Volker S, Santarius U. The three-dimensional structure of the surface protein of *Acidianus brierleyi* determined by electron crystallography. System Appl Microbiol 1991; 14:103-10.

52. Huber G, Spinnler C, Gambacorta A et al. *Metallosphaera sedula* gen. and sp. nov. represents a new genus of aerobic, metal-mobilizing, thermoacidophilic archaebacteria. System Appl Microbiol 1989; 12:38-47.

53. Segerer AH, Trincone A, Gahrtz M et al. *Stygiolobus azoricus* gen. nov., sp. nov. represents a novel genus of anaerobic, extremely thermoacidophilic archaebacteria of the order *Sulfolobales*. Int J Syst Bacteriol 1991; 41:495-501.

54. Xu Y, Li Y, Cai W et al. The biochemical properties of the cell envelope of *Sulfosphaerellus thermoacidophilum*. Acta Microbiol Sinica 1988; 28:221-25.

55. Udey LR, Fryer JL. Immunization of fish with bacterins of *Aeromonas salmonicida*. Marine Fisheries Rev 1978; 40:12-17.

56. Ishiguro EE, Kay WW, Ainsworth T et al. Loss of virulence during culture of *Aeromonas salmonicida* at high temperature. J Bacteriol 1981; 148:333-40.

57. Evenberg D, Lugtenberg B. Cell surface of the fish pathogenic bacterium *Aeromonas salmonicida*. II. Purification and characterization of a major cell envelope protein related to autoagglutination, adhesion and virulence. Biochim Biophys Acta 1982; 684:249-54.

58. Kay WW, Trust TJ. Form and functions of the regular surface array (S-layer) of *Aeromonas salmonicida*. Experientia 1991; 47:412-14.

59. Kokka RP, Vedros NA, Janda JM. Electrophoretic analysis of the surface components of autoagglutinating surface array protein-positive and -negative *Aeromonas hydrophila* and *Aeromonas sobria*. J Clin Microbiol 1990; 28:2240-47.

60. Sakata T, Shimojo T. Surface structure and pathogenicity of *Aeromonas hydrophila* strains isolated from diseased and healthy fish. Memoirs of the Faculty of Fisheries Kagoshima University 1991; 40:47-58.

61. Garduño RA, Thornton JC, Kay WW. *Aeromonas salmonicida* grown in vivo. Infect Immun 1993; 61:3854-62.

62. Chu S, Cavaignac S, Feutrier J et al. Structure of the tetragonal surface virulence array protein and gene of *Aeromonas salmonicida*. J Biol Chem 1991; 266:15258-65.

63. Gustafson CE, Chu S, Trust TJ. Mutagenesis of the paracrystalline surface protein array of *Aeromonas salmonicida* by endogenous insertion elements. J Mol Biol 1994; 237:452-63.

64. Noonan B, Trust TJ. Molecular analysis of an A-protein secretion mutant of *Aeromonas salmonicida* reveals a surface layer-specific protein secretion pathway. J Mol Biol 1995; 248:316-27.

65. Chart H, Shaw DH, Ishiguro EE et al. Structural and immunochemical homogeneity of *Aeromonas salmonicida* lipopolysaccharide. J Bacteriol 1984; 158:16-22.

66. Dooley JSG, Lallier R, Shaw DH et al. Electrophoretic and immunochemical analyses of the lipopolysaccharides from various strains of *Aeromonas hydrophila*. J Bacteriol 1985; 164:263-69.

67. Dooley JSG, Engelhardt H, Baumeister W et al. Three-dimensional structure of an open form of the surface layer from the fish pathogen *Aeromonas salmonicida*. J Bacteriol 1989; 171:190-97.

68. Dooley JSG, Trust TJ. Surface protein composition of *Aeromonas hydrophila* strains virulent for fish: identification of a surface array protein. J Bacteriol 1988; 170:499-506.

69. Belland RJ, Trust TJ. Synthesis, export, and assembly of *Aeromonas salmonicida* A-layer analysed by transposon mutagenesis. J Bacteriol 1985; 163:877-81.

70. Thomas SR, Trust TJ. Tyrosine phosphorylation of the tetragonal paracrystalline array of *Aeromonas hydrophila*: molecular cloning and high-level expression of the S-layer protein gene. J Mol Biol 1995; 245:568-81.

71. Lechner J, Wieland F. Structure and biosynthesis of prokaryotic glycoproteins. Annu Rev Biochem 1989; 58:173-94.

72. Kokka RP, Vedros NA, Janda JM. Immunochemical analysis and possible biological role of an *Aeromonas hydrophila* surface array protein in septicaemia. J Gen Microbiol 1992; 138:1229-36.

73. Kostrzynska M, Dooley JSG, Shimojo T et al. Antigenic diversity of the S-layer proteins from pathogenic strains of *Aeromonas hydrophila* and *Aeromonas veronii* biotype sobria. J Bacteriol 1992; 174:40-47.

74. Pei Z, Ellison III RT, Lewis RV et al. Purification and characterization of a family of high molecular weight surface-array proteins from *Campylobacter fetus*. J Biol Chem 1988; 263:6414-20.

75. Blaser MJ, Pei Z. Pathogenesis of *Campylobacter fetus* infections: critical role of high-molecular-weight S-layer proteins in virulence. J Infect Dis 1993; 167:372-77.

76. Kaneko T. Analysis of cell surface antigens of *Campylobacter rectus*. Bull Tokyo Dental College 1992; 33:171-85.

77. Grollier G, Burucoa C, Ricco JB et al. Isolation and immunogenicity of *Campylobacter fetus* subsp. *fetus* from an abdominal aortic aneurysm. Eur J Clin Microbiol Infect Dis 1993; 12:847-49.

78. Tummuru MKR, Blaser MJ. Rearrangement of *sapA* homologs with conserved and variable regions in *Campylobacter fetus*. Proc Natl Acad Sci USA 1993; 90:7265-69.

79. Dworkin J, Tummuru MKR, Blaser MJ. Segmental conservation of *sap*A sequences in type B *Campylobacter fetus* cells. J Biol Chem 1995; 270:15093-101.

80. Garcia MM, Lutze-Wallace CL, Denes AS et al. Protein shift and antigenic variation in the S-layer of *Campylobacter fetus* subsp. *veneralis* during bovine infection accompanied by genomic rearrangement of *sapA* homologs. J Bacteriol 1995; 177:1976-80.

81. Dasch GA. Isolation of species-specific protein antigens of *Rickettsia typhi* and *Rickettsia prowazekii* for immunodiagnosis and immunoprophylaxis. J Clin Microbiol 1981; 14:333-41.

82. Ching W-M, Dasch GA, Carl M et al. Structural analyses of the 120-kDa serotype protein antigens of typhus group rickettsiae. Comparison with other S-layer proteins. Ann New York Acad Sci 1990; 590:334-51.

83. Carl M, Dobson ME, Ching W-M et al. Characterization of the gene encoding the protective paracrystalline-surface-layer protein of *Ricksettsia prowazekii*: presence of a truncated identical homolog in *Ricksettsia typhi*. Proc Natl Acad Sci USA 1990; 87:8237-41.

84. Ching W-M, Carl M, Dasch GA. Mapping of monoclonal antibody binding sites on CNBr fragments of the S-layer protein antigens of *Ricksettsia typhi* and *Ricksettsia prowazekii*. Mol Immunol 1992; 29:95-105.

85. Gilmore Jr RD, Joste N, McDonald GA. Cloning, expression and sequence analysis of the gene encoding the 120kD surface-exposed protein of *Rickettsia rickettsii*. Mol Microbiol 1989; 3:1579-86.

86. Kawata T, Takeoka A, Takumi K et al. Demonstration and preliminary characterization of a regular array in the cell wall of *Clostridium difficile*. FEMS Microbiol Lett 1984; 24:323-28.

87. Takumi K, Koga T, Oka T et al. Self-assembly, adhesion and chemical properties of tetragonaly arrayed S-layer proteins of *Clostridium*. J Gen Appl Microbiol 1991; 37:455-65.

88. Takumi K, Endo Y, Koga T et al. In vitro self-assembly of the S-layer subunits from *Clostridium difficile* GAI 0714 into tetragonal arrays. Tokushima J Exp Med 1992; 39:95-100.

89. Hagiya H, Oka T, Tsuji H et al. The S-layer composed of two different protein subunits from *Clostridium difficile* GAI 1152: a simple purification method and characterization. J Gen Appl Microbiol 1992; 38:63-74.

90. Takumi K, Susami Y, Takeoka A et al. S-layer protein of *Clostridium tetani*: purification and properties. Microbiol Immunol 1991; 35:569-76.

91. Takumi K, Ichiyanagi S, Endo Y et al. Characterization, self-assembly and reattachment of S-layer from *Clostridium botulinum* type E Saroma. Tokushima J Exp Med 1992; 39:101-7.

92. Etienne-Toumelin I, Sirard J-C, Duflot E et al. Characterization of the *Bacillus anthracis* S-layer: cloning and sequencing of the structural gene. J Bacteriol 1995; 177:614-20.

93. Farchaus JW, Ribot WJ, Downs MB et al. Purification and characterization of the major surface array protein from the avirulent *Bacillus anthracis* Delta Sterne-1. J Bacteriol 1995; 177:2481-89.

94. Luckevich MD, Beveridge TJ. Characterization of a dynamic S-layer on *Bacillus thuringiensis*. J Bacteriol 1989; 171:6656-67.

95. Kobayashi Y, Ohta H, Kokeguchi S et al. Antigenic properties of *Campylobacter rectus* (*Wolinella recta*) major S-layer proteins. FEMS Microbiol Lett 1993; 108:275-80.

96. Lortal S, van Heijenoort J, Gruber K et al. S-layer of *Lactobacillus helveticus* ATCC 12046: isolation, chemical characterization and reformation after extraction with lithium chloride. J Gen Microbiol 1992; 138:611-18.

97. Masuda K. Heterogeneity of S-layer proteins of *Lactobacillus acidophilus* strains. Microbiol Immunol 1992; 36:297-301.

98. Toba T, Virkola R, Westerlund B et al. A collagen-binding S-layer protein in *Lactobacillus crispatus*. Appl Environ Microbiol 1995; 61:2467-71.

99. Lortal S, Rouault A, Cesselin B et al. Paracrystalline surface layers of dairy propionibacteria. Appl Environ Bacteriol 1993; 59:2369-74.

100. Sára M, Sleytr UB. Comparative studies of S-layer proteins from *Bacillus stearothermophilus* strains expressed during growth in continuous culture under oxygen-limited and non-oxygen-limited conditions. J Bacteriol 1994; 176:7182-89.

101. MacRae JD, Smit J. Characterization of caulobacters isolated from waste water treatment systems. Appl Environ Bacteriol 1991; 57:751-58.

102. Walker SG, Smith SH, Smit J. Isolation and comparison of the paracrystalline surface layer proteins of freshwater caulobacters. J Bacteriol 1992; 174:1783-92.

103. Kawata T, Masuda K, Yoshino K et al. Regular array in the cell wall of *Lactobacillus fermenti* as revealed by freeze-etching and negative staining. Jpn J Microbiol 1974; 18:469-76.

104. Lortal S, Rousseau M, Boyaval, P et al. Cell wall and autolytic system of *Lactobacillus helveticus* ATCC 12046. J Gen Microbiol 1991; 137:549-59.

105. Lortal S. Crystalline surface-layers of the genus Lactobacillus. In: Beveridge TJ, Koval SF, eds. Advances in Bacterial Paracrystalline Surface Layers. New York: Plenum, 1993:57-65.

106. Mozes N, Lortal S. X-ray photoelectron spectroscopy and biochemical ana lysis of the surface of *Lactobacillus helveticus* ATCC 12046. Microbiology 1995; 141:11-19.

107. Smit J. Protein surface layers of bacteria. In: Inouye M, ed. Bacterial Outer Membranes as Model Systems. New York: Wiley, 1987:343-76.

108. Smit J, Engelhardt H, Volker S et al. The S-layer of *Caulobacter crescentus*: three-dimensional image reconstruction and structure analysis by electron microscopy. J Bacteriol 1992; 174:6527-38.

109. Gilchrist A, Fisher JA, Smit J. Nucleotide sequence analysis of the gene encoding the *Caulobacter crescentus* paracrystalline surface layer protein. Can J Microbiol 1992; 38:193-202.

110. Takagi H, Shida O, Kadowaki K et al. Characterization of *Bacillus brevis* with descriptions of *Bacillus migulanus* sp. nov., *Bacillus choshinensis* sp. nov., *Bacillus parabrevis* sp. nov., and *Bacillus galactophilus* sp. nov. Int J Syst Bacteriol 1993; 43:221-31.

111. Shida O, Takagi H, Kadowaki K et al. *Bacillus aneurinolyticus* sp. nov., nom. rev. Int J Syst Bacteriol 1994; 44:143-50.

112. Shida O, Takagi H, Kadowaki K et al. Proposal of *Bacillus reuszeri* sp. nov., *Bacillus formosus* sp. nov., nom. rev., and *Bacillus borstelensis* sp. nov., nom. rev. Int J Syst Bacteriol 1995; 45:93-100.

113. Karnauchow TM, Koval SF, Jarrell KF. Isolation and characterization of three thermophilic anaerobes from a St. Lucia hot spring. System Appl Microbiol 1992; 15:296-310.

114. Sleytr UB, Sára M, Küpcü Z et al. Structural and chemical characterization of S-layers of selected strains of *Bacillus stearothermophilus* and *Desulfotomaculum nigrificans*. Arch Microbiol 1986; 146:19-24.

115. Lis H, Sharon N. Protein glycosylation. Structural and functional aspects. Eur J Biochem 1993; 218:1-27.

116. Bahl OP. An introduction to glycoproteins. In Allen HJ, Kisailus EC, eds. Glycoconjugates. Composition, Structure, and Function. New York: Dekker, 1992:1-13.

117. Sumper M, Wieland FT. Bacterial glycoproteins. In Montreuil J, Vliegenthart JFG, Schachter H, eds. Glycoproteins. Amsterdam: Elsevier, 1995:455-73.

118. Mescher MF, and Strominger JL. Purification and characterization of a prokaryotic glycoprotein from the cell envelope of *Halobacterium salinarium*. J Biol Chem 1976; 251:2005-14.

119. Sumper M. Halobacterial glycoprotein biosynthesis. Biochim Biophys Acta 1987; 906:69-79.

120. Sumper M, Berg E, Mengele R et al. Primary structure and glycosylation of the S-layer protein of *Haloferax volcanii*. J Bacteriol 1990; 172:7111-18.

121. Mengele R, Sumper M. Drastic differences in glycosylation of related S-layer glycoproteins from moderate and extreme halophiles. J Biol Chem 1992; 267:8182-85.

122. Zaccai G, Cendrin F, Haik Y et al. Stabilization of halophilic malate dehydrogenase. J Mol Biol 1989; 208:491-500.

123. Zhu BCR, Drake RR, Schweingruber H et al. Inhibition of glycosylation by amphomycin and sugar nucleotide analogs PP36 and PP55 indicates that *Haloferax volcanii* β-glucosylates both glycoproteins and glycolipids through lipid-linked intermediates: evidence for three novel glycoproteins and a novel sulfated dihexosyl-archaeol glycolipid. Arch Biochem Biophys 1995; 319:355-64.

124. Nishiyama Y, Takashina T, Grant WD et al. Ultrastructure of the cell wall of the triangular halophilic archaebacterium *Haloarcula japonica* strain TR-1. FEMS Microbiol Lett 1992; 99:43-48.

125. Nakamura S, Mizutani S, Wakai H et al. Purification and partial characterization of cell surface glycoprotein from extremely halophilic archaeon *Haloruca japonica* strain TR-1. Biotechnol Lett 1995; 17:705-6.

126. Bröckl G, Behr M, Fabry S et al. Analysis and nucleotide sequence of the genes encoding the surface-layer glycoproteins of the hyperthermophilic methanogens *Methanothermus fervidus* and *Methanothermus sociabilis*. Eur J Biochem 1991; 199:147-52.

127. Hartmann E, König H. Uridine and dolichyl diphosphate activated oligo saccharides are intermediates in the biosynthesis of the S-layer glycoprotein of *Methanothermus fervidus*. Arch Microbiol 1989; 151:274-81.

128. Bayley DP, Koval SF. Membrane association and isolation of the S-layer protein of *Methanoculleus marisnigri*. Can J Microbiol 1994; 40:237-41.

129. Cheong G-W, Cejka Z, Peters J et al. The surface protein layer of *Methanoplanus limicola*: three-dimensional structure and chemical characterization. System Appl Microbiol 1991; 14:209-17.

130. Pley U, Schipka J, Gambacorta A et al. *Pyrodictium abyssi* sp. nov. represents a novel heterotrophic marine archaeal hyperthermophile growing at 110°C. System Appl Microbiol 1991; 14:245-53.

131. Bashkatova NA, Severina LO, Golovacheva RS et al. Surface layers of extremely thermoacidophilic archaebacteria of the genus *Sulfolobus*. Mikrobiologiya 1991; 60:90-94.

132. Sleytr UB, Thorne KJI. Chemical characterization of the regularly arrayed surface layers of *Clostridium thermosaccharolyticum* and *Clostridium thermohydrosulfuricum*. J Bacteriol 1976; 126:377-83.

133. Severina LO, Senyushkin AA, Karavaiko GI. Ultrastructure and chemical composition of the S-layer of *Sulfobacillus thermo-sulfidooxidans*. Dokl Akad Nauk 1993; 328:633-36.

134. Peyret JL, Bayan N, Joliff G et al. Characterization of the *cspB* gene encoding PS2, an ordered surface-layer protein in *Corynebacterium glutamicum*. Mol Microbiol 1993; 9:97-109.

135. Smith SH, Murray RGE. The structure and associations of the double S-layer on the cell wall of *Aquaspirillum sinuosum*. Can J Microbiol 1990; 36:327-35.

136. Knirel YA, Kochetkov NK. The structure of lipopolysaccharides of Gram-negative bacteria. III. The structure of O-antigens. Biochemistry (Moscow) 1994; 59:1325-83.

137. Christian R, Schulz G, Unger FM et al. Structure of a rhamnan from the surface layer glycoprotein of *Bacillus stearothermophilus* strain NRS 2004/3a. Carbohydr Res 1986; 150:265-72.

138. Messner P, Sleytr UB, Christian R et al. Isolation and structure determination of a diacetamidodideoxyuronic acid-containing glycan chain from the S-layer glycoprotein of *Bacillus stearothermophilus* NRS 2004/3a. Carbohydr Res 1987; 168:211-18.

139. Messner P, Sleytr UB. Asparaginyl-rhamnose: a novel type of protein-carbohydrate linkage in a eubacterial surface-layer glycoprotein. FEBS Lett 1988; 228:317-20.

140. Kosma P, Wugeditsch T, Christian R et al. Glycan structure of a heptose-containing S-layer glycoprotein of *Bacillus thermoaerophilus*. Glycobiology; in press.
141. Altman E, Brisson J-R, Messner P et al. Chemical characterization of the regularly arranged surface layer glycoprotein of *Bacillus alvei* CCM 2051. Biochem Cell Biol 1991; 69:72-78.
142. Christian R, Messner P, Weiner C et al. Structure of a glycan from the surface-layer glycoprotein of *Clostridium thermohydrosulfuricum* strain L111-69. Carbohydr Res 1988; 176:160-63.
143. Messner P, Christian R, Kolbe J et al. Analysis of a novel linkage unit of O-linked carbohydrates from the crystalline surface layer glycoprotein of *Clostridium thermohydrosulfuricum* S102-70. J Bacteriol 1992; 174:2236-40.
144. Christian R, Schulz G, Schuster-Kolbe J et al. Complete structure of the tyrosine-linked saccharide moiety from the surface layer glycoprotein of *Clostridium thermohydrosulfuricum* S102-70. J Bacteriol 1993; 175:1250-56.
145. Altman E, Brisson J-R, Gagné SM et al. Structure of the glycan chain from the surface layer glycoprotein of *Clostridium thermohydrosulfuricum* L77-66. Biochim Biophys Acta 1992; 1117:71-77.
146. Peters J, Rudolf S, Oschkinat H et al. Evidence for tyrosine-linked glycosaminoglycan in a bacterial surface protein. Biol Chem Hoppe-Seyler 1992; 373:171-76.
147. Lupas A, Engelhardt H, Peters H et al. Domain structure of the *Acetogenium kivui* surface layer revealed by electron crystallography and sequence analysis. J Bacteriol 1994; 176:1224-33.
148. Altman E, Brisson J-R, Messner P et al. Chemical characterization of the regularly arranged surface layer glycoprotein of *Clostridium thermosaccharolyticum* D120-70. Eur J Biochem 1990; 188:73-82.
149. Altman E, Schäffer C, Brisson J-R et al. Characterization of the glycan structure of a major glycopeptide from the surface layer glycoprotein of *Clostridium thermosaccharolyticum* E207-71. Eur J Biochem 1995; 229:308-15.
150. Messner P, Bock K, Christian R et al. Characterization of the surface layer glycoprotein of *Clostridium symbiosum* HB25. J Bacteriol 1990; 172:2576-83.
151. Möschl A, Schäffer C, Sleytr UB et al. Characterization of the S-layer glycoproteins of two lactobacilli. In: Beveridge TJ, Koval SF, eds. Advances in Bacterial Paracrystalline Surface Layers. New York: Plenum, 1993:281-84.
152. Pellerin P, Fournet B, Debeire P. Evidence for the glycoprotein nature of the cell sheath of *Methanosaeta*-like cells in the culture of *Methanothrix soehngenii* strain FE. Can J Microbiol 1990; 36:631-36.

153. Sára M, Küpcü S, Sleytr UB. Localization of the carbohydrate residue of the S-layer glycoprotein from *Clostridium thermohydrosulfuricum* L111-69. Arch Microbiol 1989; 151:416-20.

154. Hartmann E, Messner P, Allmaier G et al. Proposed pathway for biosynthesis of the S-layer glycoprotein of *Bacillus alvei*. J Bacteriol 1993; 175:4515-19.

155. Schäffer C, Neuninger C, Wugeditsch T et al. Are S-layer glycopoteins and lipopolysaccharides related? In: Tomasz A, ed. Microbial Drug Resistance: Mechanisms, Epidemiology and Disease. Larchmont, NY: Liebert, in press.

156. Baumeister W, Engelhardt H. Three-dimensional structure of bacterial surface layers. In: Harris JR, Horne RW, eds. Electron Microscopy of Proteins, Vol. 6, Membraneous Structure. London: Academic Press, 1987:109-54.

157. Hovmöller S, Sjögren A, Wang DN. The structure of crystalline bacterial surface layers. Prog Biophys Mol Biol 1988; 51:131-63.

158. Blaser MJ, Gotschlich EC. Surface array protein of *Campylobacter fetus*. J Biol Chem 1990; 265:14529-35.

159. Sanderock LE, MacLeod AM, Ong E et al. Non-S-layer glycoproteins in eubacteria. FEMS Microbiol Lett 1994; 118:1-8.

CHAPTER 4

ANALYSIS OF S-LAYER PROTEINS AND GENES

Beatrix Kuen, Werner Lubitz

4.1. INTRODUCTION

Surface layers (S-layers) are regularly ordered proteins found as the outermost cell envelope component of many bacteria (see chapter 2).[1,2,3] The majority of S-layers are composed of single protein or glycoprotein species with a molecular weight varying between 40 kDa and 200 kDa. These proteins are highly expressed proteins and in most cases they are among the major protein species produced by the cell.

Although S-layer proteins have been reported on hundreds of different species on nearly every taxonomic group of bacteria, up to now only 27 S-layer encoding genes have been cloned and sequenced (see Table 4.1). While S-layer proteins constitute an important class of secreted proteins, the regulation of S-layer protein synthesis, their translocation across the cell envelope and information about S-layer domains essential for intra- and/or intermolecular interactions is still poor. This is mainly the result of difficulties commonly encountered in cloning S-layer protein genes in a form where they stably display high level expression in the cloning host and where the formation of regular S-layer arrays is possible.

Crystalline Bacterial Cell Surface Proteins, edited by Uwe B. Sleytr, Paul Messner, Dietmar Pum, Margit Sára. © 1996 R.G. Landes Company.

Assuming that an S-layer protein gene can be stably cloned in an homologous or heterologous system, molecular biology and site directed mutagenesis techniques open the possibility for structural/ functional analyses. Recently, heterologous expression and self-assembly of an S-layer protein has been achieved for the first time in *Escherichia coli* with the S-layer protein SbsA of *Bacillus*

Table 4.1. S-layer genes sequenced, aa number deduced and lattice type of corresponding proteins

Organism	S-layer gene	aa residues of proteins	Signal sequence aa	Lattice type	No. of sulfur aa (Cys+Met)	Ref.
Acetogenum kivui [a]	*slp*	736	26	H	0+6	4
Aeromonas hydrophila	*ahs*	448	19	S	0+5	5
Aeromonas salmonicida	*vap*	481	21	S	0+3	6
Bacillus anthracis	*sap*	785	29	H	0+10	7
Bacillus brevis	*mwp*	1053	23	H	0+19	8
Bacillus brevis	*owp*	1004	24	H	0+6	9
Bacillus brevis	*HWP gene*	1087	23	H	1+19	10
Bacillus sphaericus	*gene 125*	1176	30	S	0+6	11
Bacillus sphaericus	*gene 80*	745	No	n.e.	0+6	11
Bacillus stearothermophilus	*sbsA*	1199	30	H	0+3	12
Campylobacter fetus	*sapA*	933	No	H	1+15	13
Campylobacter fetus	*sapA1*	920	No	P	n.d.	14
Campylobacter fetus	*sapA2*	1109	No	P	0+6	15
Caulobacter crescentus	*rsaA*	1026	No	H	0+4	16
Corynebacterium glutamicum	*cspB*	510	30	P	0+1	17
Deinococcus radiodurans	*HPI gene*	1036	31	H	6+6	18
Halobacterium halobium	*csg*	852	34	H	0+9	19
Haloferax volcanii	*S-l. gene* [b]	828	34	H	0+7	20
Lactobacillus brevis	*S-l. gene* [b]	465	30	S	0+5	21
Lactobacillus acidophilus	*slpA*	444	24	P	0+2	22
Methanothermus fervidus	*slgA* [c]	593	22	H	8+9	23
Methanothermus sociabilis	*slgA* [c]	593	22	H	8+11	23
Methanococcus voltae	*sla*	565	12	H	0+12	24
Rickettsia prowazekii	*spaP*	1612	No	S	3+16	25
Rickettsia rickettsii	*p120*	1299	No	S	2+10	26
Rickettsia typhii	*slpT*	1645	32	S	0+16	27
Thermus thermophilus	*slpA*	917	27	S	0+10	28

Abreviations: aa, amino acids. [a] reclassified: *Thermoanaerobacter kivui*. [b] S-l.gene: S-layer gene. [c] The nucleotide sequences of these genes differ at only nine positions, resulting in three amino acid differences. The type of S-layer lattice is abbreviated: H, hexagonal; S, square; P, periodic structure not further characterized; n.e., not expressed.

stearothermophilus PV72. This achievement opens the way for structural investigations of the protein as well as of construction of recombinant S-layer proteins for various applications in molecular biotechnology (Kuen B, Sára M, Lubitz W. Heterologous expression and self-assembly of the S-layer protein SbsA of *Bacillus stearothermophilus* in *Escherichia coli*. Submitted 1995).

With these results in mind, in the following chapter we will discuss similarities and common characteristics of currently known S-layer genes, with an emphasis on possibly important sequence features which may suggest a starting point for future work.

4.2. PROMOTER STRUCTURES OF S-LAYER GENES

S-layer proteins are predominant proteins produced by bacterial cells. It has been calculated that approximately 5×10^5 S-layer protein monomers are needed to cover a bacterial cell, and for bacteria growing with a generation time of 20 minutes, this requires a production rate of approximately 500 molecules per second (see chapter 2).[29] It is obvious that S-layer protein expression must be very efficient and that regulatory circuits are necessary to ensure its synthesis is coordinated with cell growth. However, of the 27 S-layer genes sequenced (Table 4.1) only eight mRNAs have been experimentally analyzed by primer extension and/or S1 mapping for their transcriptional start points (Table 4.2). As shown for these S-layer protein genes, the transcripts were found to be monocistronic, with the exception of the polycistronic mRNA of *Bacillus brevis* 47 encoding the S-layer genes *mwp* and *owp*. *Bacillus brevis* 47 produces a double protein layer, consisting of the middle wall layer (MWP) and the outer wall layer (OWP). The structural genes encoding MWP and OWP are in tandem and constitute a cotranscriptional unit.[30]

4.2.1. TRANSCRIPTION AND EXPRESSION OF DIFFERENT S-LAYER GENES

A comparison of all experimentally determined promoter regions of S-layer genes available is given in Table 4.2. The alignment has been made by using the -35 and -10 consensus regions of *E. coli* promoters as well as the archaeobacterial A and B box sequences.

Primer extension analysis of S-layer mRNA of *B. brevis* revealed the existence of five (P1-P5; Table 4.2) tandemly arranged promoters in the 5′ region of the *cwp* (cell wall protein) operon. The -35 and -10 regions of the P1 and P3 promoters resemble the consensus sequence recognized by the sigma-43 type RNA polymerase of *Bacillus subtilis*. The P2 promoter shows only homology to the consensus sequence in the -10 region. The P2 promoter is used constitutively in *B. brevis* 47 at all stages of growth, whereas

Table 4.2. Comparison of putative promoter regions of various S-layer genes

Gene	Promoter	Promoter region	L-RNA
		-35 region (Consensus *E. coli*) -10 region	
		TTGACA TATAAT	
	P2	TAAATATAGAAAAATACTAAAAATTTAGTATTA	170
vapA	P1	TTGACATGCCTGAGTCTGATGGCTAAGAA	181
	P2	CGGATAGGTTCAACCCTATTTGTATATAAT	n.i.
sapA	P	TTGAATTTTATAAAAAATTATGTTATAAT	114
rsaA	P	TTGTCGACGTATGACGTTTGCTCTATA	60
S-1.gene	P1	TTGTATCTTCCTTAAGGAAATCGCTATACT	127
	P2	TTAACAAAAGCGCTAACTTCGGTTATACT	43
slpA	P	TTGACAAGGGCGCGTGAGGTTTTTACGAT	127
cwp	P1	GTGACAGCCCGCCATATGTCCCCTATAAT	299
	P2	CCGCAACTTTTGATTCGCTCAGGCGTTTAAT	257
	P2*	TTGATTCGCTCAAAAGGCGTTTAATAGGATGT	247
	P3	GATTCACGAATTCTAGCAGTTGTGTTACACT	176
	P4	TACACAATACTGAATATACTAGAGATTTTTAAC	130
	P5	TATACTAGAGATTTTTAACACAAAAAGCGAG	117
		A-Box (Consensus Archaeobacteria) B-Box	
		TTTAWATA WTGM	
sla	P1	ATCATATATTATTTATACGAGTATTGACTACTCGTT	244
	P2	TAAATATAGAAAAATACTAAAAATTTAGTATTA	170
	P3	TTTAAAAGTTTAACAACTTATTGATTACTACCGGA	110
slgA	P1	TATATATATGAAAGAAACTTTGTATTTCTAATGT	32

Origins of S-layer genes: *vapA*, Aeromonas salmonicida;[6] *sapA* , Campylobacter fetus;[13] *rsaA*, Caulobacter crescentus;[16] *S-l. gene*, Lactobacillus brevis;[21] *slpA*, Thermus thermophilus,[28] *cwp*, Bacillus brevis;[30] P1 to P3 were identified and P2*, P4 and P5 are putative promoters; *sla*, Methanococcus voltae;[24] *slgA*, Methanothermus fervidus and Methanothermus sociabilis.[23] Possible eubacterial -35 and -10 regions and archaeobacterial A- and B-Boxes, repectively, are underlined. P, promoter. n.i., not identified. L-RNA, length of leader RNA in bases. W=A or T; M=A or C .

promoter P3 is used only during the exponential phase of growth. Thus, promoter P2 might be responsible for the constitutive synthesis and secretion of the cell wall proteins into the medium during the stationary phase of growth.[30] The transcriptional activities of promoters P4 and P5 with regard to the cell cycle of *B. brevis* 47 has not yet been elucidated.

In the case of *Lactobacillus brevis,* two promoters have been found which appear to be equally active during exponential growth phase.[21]

In *Aeromonas salmonicida* two promoters (P1, P2) have been proposed based on computer analysis, however, only promoter P1 has been mapped by primer extension analysis. This promoter appears to be sigma-70 type RNA polymerase dependent, having a characteristic -35 consensus sequence, and a related -10 motif.[31] The *vapA* transcript from promoter P1 is the major species present and peaks in the mid-log growth phase of *A. salmonicida.* The predicted P2 sequence serves as a promoter, albeit its strength seems to be poor. During culture of *Aeromonas salmonicida,* the expression of *vapA* has been observed to be subject to regulation. Further studies, identified the regulatory factor AbcA.[32] AbcA is a bifunctional protein consisting of an ATP-binding domain, which is required for the synthesis for the O-polysaccharide side chains[33] and a leucine zipper domain which influences expression of the S-layer gene *vapA*. It is not clear whether or not AbcA is the sole factor affecting *vapA* expression.[34]

For the S-layer encoding structural gene (*sla*) of the archaeobacterium *Methanococcus voltae,* it was shown that *sla* is initiated at three distinct promoter sites. Upstream from each transcription start point is a region with similarity to the *Box* A consensus sequence observed in archaeobacterial promoters.[35] In two (P1 and P3) of the three promoters, two *Box* A sequences were present in tandem. This tandem arrangement could be responsible for high level of *sla* expression. Presumptive archaeobacterial *Box* B signatures were also identified in all three promoters.[36]

Only one transcriptional start point has been mapped for the S-layer glycoprotein encoding gene of the archaeobacterium *Halobacterium halobium,*[19] as well as for the S-layer protein encoding genes of the archaeobacterial organisms *Methanothermus fervidus* and *Methanothermus sociabilis*.[23] The transcription start sites in both *Methanothermus* species were localized 33 bp upstream of the predicted translation initiation codon at the guanosine residue in an ATGT sequence, which resembles the *Box* B consensus sequence of promoters of archaeobacteria. A nucleotide sequence conforming to the *Box* A promoter consensus sequence of archaeobacteria is located 24 bp upstream of the *Box* B sequence.

The S-layer gene of *Thermus thermophilus* HB8 is transcribed as a monocistronic unit with a 127 bp leader mRNA.[28] The -35 region of *slpA* is very similar to the consensus for *E. coli*, while the -10 region is less similar. In addition, it has been demonstrated that the *slpA* promoter is also active in *E. coli*, supporting the apparent sequence homologies.

Primer extension analysis of the *Campylobacter fetus* S-layer gene (*sapA*) mapped one transcription initiation site at position -114 upstream from the translational start site.[37] The *sapA* transcription initiation site is spaced seven nucleotides downstream from the hexamer TATAAT, which exhibits the most highly conserved nucleotides for *E. coli* -10 promoter elements.[38]

The site of transcription initiation for the S-layer gene (*rsaA*) of *Caulobacter crescentus* has been determined by S1 nuclease mapping 60 bp upstream of the translation initiation codon.[36] Upstream of the *rsa* transcription start site at -10 and -35 bp a consensus promoter sequence homologous to *E. coli* has been determined. The *rsaA* promoter seems not to be developmentally regulated as transcription is constant during the cell cycle.[39]

4.2.2. LEADER mRNA OF S-LAYER PROTEIN GENES AND TRANSCRIPT STABILITY

As can be seen in Table 4.2, most S-layer protein mRNAs have long leader sequences. It has been suggested that the long mRNA leader sequences may have an important role in the regulation of S-layer gene expression, since extended mRNA leader sequences are normally associated with highly expressed genes, e.g., rRNA[40] operons and the *ompA* gene.[41]

In *Aeromonas salmonicida*, secondary structural predictions of the *vapA* transcript have suggested that the leader mRNA sequence contains two stem-loop structures. Stem-loop structures in the 5′region of mRNA can function as attenuators as with the *ompA* transcript in *E. coli*.[42,43] On the other hand, stem-loop structures are also known to inhibit the initial cleavages leading to exonuclease degradation of mRNAs and thus are features of mRNAs with long intracellular half-lives.[44] For example, the *ompA* transcript has a half life of approximately 17 minutes,[41] in contrast to the average half-life of most prokaryotic mRNAs which is approximately three minutes.[45,46]

In the 127 nt long *slpA* leader mRNA of *Thermus thermophilus,* a 13-base long inverted repeat has been identified which most probably functions to protect the 5′ terminus from exonuclease digestion and thereby increases the half-life of the mRNA.[28]

In the *Campylobacter fetus sapA* mRNA leader transcript, a thrice-repeated pentameric unit (ATTTT) has been observed, however, the significance of this repeated sequence has still to be determined.[37]

The half-life time of the *vapA* mRNA of *Aeromonas salmonicida* is 22 minutes in cells growing at 15°C,[31] and for the *Caulobacter crescentus rsaA* mRNA the half-life time was determined to be 15 minutes.[39] The high stability of S-layer mRNAs obviously contributes to the high level production of S-layer proteins. Growth rate dependent regulation of gene expression as well as the protein stability itself and turnover rates in S-layer sheet formation and/or recycling of monomer or protomer units are only poorly understood.

4.2.3. IMPAIRMENT OF 5′UPSTREAM GENE SEQUENCES IN S-LAYER DEFICIENT MUTANT STRAINS

Southern hybridization analysis of S-layer deficient mutant strains from the organisms *Aeromonas salmonicida,*[47] *Bacillus stearothermophilus* (Sara M, Kuen B, Mayer HF et al. Dynamics in oxygen-induced changes in S-layer protein synthesis and cell wall composition in continuous culture from *Bacillus stearothermophilus* PV72 and the S-layer-deficient variant T5. Submitted 1995), *Campylobacter fetus*[37] and *Campylobacter rectus*[48] indicate that such mutant strains still carry S-layer gene sequences within their chromosomes. Compared with the corresponding S-layer encoding wild-type DNA sequences, the mutant strains differed in the

5'upstream region of the S-layer genes. In the case of *A. salmonicida*, loss of S-layer expression results from growth above 25°C, and was accompanied by genetic rearrangements in which 5' sequences of the gene were lost by deletion.

For *B. stearothermophilus* PV72, hybridization and PCR analysis suggested that in the S-layer deficient mutant strain T5 the *sbsA* 5'region has a deletion of approximately 800 bases compared to the wild-type strain (Kuen B, Scholz H; unpublished results).

Since in *C. rectus* no significant differences were found in the hybridization patterns of the 5'upstream region of S+ and S- strains, it was suggested that the loss of the S-layer in the mutant strain depends on minor sequence changes.[48]

In the case of *Campylobacter fetus* hybridization analyses with a DNA probe, derived from the *sapA* promoter region of the wild-type strain 23D, revealed no signal with the S-layer deficient strain 23B and Northern RNA blot analyses showed no *sapA* mRNA in strain 23B. These data indicated that the lack of S-layer protein expression in spontaneous mutant strains is caused by the deletion of promoter sequences.[37]

4.3. SECRETION OF S-LAYER PROTEINS

The mechanisms by which S-layer proteins are secreted from the cell and assembled on the cell surface have not been characterized, in spite of the fact that S-layer proteins are a major class of secreted proteins.

All S-layer proteins with detailed information of their sequences are preceded by a signal sequence with exception of the S-layer proteins of *C. fetus*,[13] *C. crescentus*,[16] *R. prowazekii*[25] and *R. typhii*[27] (see Table 4.1).

In case of *C. crescentus* a series of gene fusions were created to study the mechanism of secretion of the S-layer protein RsaA. The results of this studies revealed that large portions of the S-layer protein N-terminus cannot mediate export of passenger proteins from the cytoplasm and that the entire native S-layer protein may be required to properly interact with the RsaA secretion machinery.[49]

Components of the extracellular secretion pathway for the translocation of the surface layer proteins through the outer membrane have been identified for the organism *A. salmonicida* (ApsE) and *A. hydrophila* (SpsD). [50,51]

The *apsE* gene is conserved among the *A. salmonicida* strains, but not in any of the S-layer producing *A. hydrophila* or *A. veronii* strains. The deduced product of the *apsE* gene exhibits homology to a number of proteins involved in extracellular protein secretion in bacteria. A conserved feature of these proteins is the presence of an ATP-binding domain. Investigations with the *apsE* mutant A449-TM1 revealed that this mutant is capable of secreting a wide range of proteins and is only impaired in its ability to secrete the S-layer proteins indicated by accumulation of S-layer proteins in the periplasm. This is strong evidence that the pathway for the S-layer protein is a specific secretion pathway and other pathway(s) must exist for the secretion of the other extracellular proteins produced by this organism. [50]

The *spsD* (S-protein secretion D) gene of *A. hydrophila* TF7 encoding a PulD homolog is located 700 bp upstream of the *ahsA* gene, the structural gene for the S-layer protein. The corresponding Protein SpsD shows homology to ExeD an other PulD homolog in *A. hydrophila*. The mechanism how PulD-like proteins function in their secretion of various substrates across membranes is unknown. Insertional inactivation of the *spsD* gene resulted in the loss of S-layer from the cell surface and periplasmic accumulation of S-layer protein but had no apparent effect on the outer-membrane-protein profile of the organism or the secretion of extracellular enzymes. This finding is consistent with the notion of S-layer producing strains having separate pathways for S-layer protein subunits, extracellular enzymes and outer membrane proteins. [51]

4.4. PROTEIN HOMOLOGY STUDIES AMONG DIFFERENT S-LAYER PROTEINS AND RELATED PROTEINS

S-layer proteins commonly contain a large portion of acidic, hydrophobic and hydroxy amino acids and are low in sulfur containing amino acids with the exception of archaeobacterial S-layer proteins (Table 4.1). Available data indicate that S-layer amino acid sequence homologies among different bacteria are extremely rare.

This notion could however, be biased by the low number of sequenced S-layer protein genes, despite the fact that S-layers are
widespread throughout the whole phylogenetic tree, including,
eubacteria and arachaeobacteria (see chapter 2). For example, the
search for the relatedness of the S-layer protein SbsA of
B. stearothermophilus PV72 with other S-layer proteins was performed using the NCBI-Blast program.[52] The best scores of SbsA
were obtained with seven out of 27 S-layer proteins and with flagellins from different bacterial species. Interestingly, five of the related
S-layer proteins belong to the genus *Bacillus* and the two others
belong to *Lactobacillus* and *Caulobacter* (Table 4.3). No homology

Table 4.3. Homology between the S-layer protein SbsA and other S-layer proteins

Organism	Protein	Residues	% Identity (a)	% Similarity (b)	Ref.
Bacillus anthracis	Sap	785	24.8%	45.4%	7
Bacillus sphaericus	125K	1176	24.2%	42.9%	11
Bacillus brevis	OWP	1004	23.6%	41.8%	9
Bacillus sphaericus	80K	745	22.8%	41.2%	11
Bacillus brevis HPD	HWP	1087	22.5%	42.4%	10
Lactobacillus acidophilus	SlpA	444	21.7%	38.7%	22
Caulobacter crescentus	RsaA	1025	21.6%	37.7%	16

Table 4.4. Homology between the S-layer protein SbsA and flagellins

Organism	Protein	Residues	% Identity (a)	% Similarity (b)	Ref.
Escherichia coli	flagellin	595	24.7%	46.1%	54
Campylobacter jejunii	flagelin B	575	24.5%	43.9%	55
Salmonella typhimurium	phase-1 i flagellin	490	23.7%	46.8%	55
Campylobacter jejunii	flagellin A	575	20.8%	44.3%	56

The alignments were performed by the GCG-program Best-Fit, using the homology algorithm of Smith and
Waterman, with a gap penalty of three. (a) Percentage of identical amino acid pairs in aligned sequences.
(b) Percentage of conservatively similar amino acid pairs in the aligned sequences.

```
         10         20         30         40         50         60         70
   MDRKKAVKLA TASAIAASAF VAANPNASEA ATDVATVVSQ AKAQFKKAYY TYSHTVTETG EFPNINDVYA  SbsA Bs
   .......... .......... .....MAQVI NTNSLSLITQ N......... .......... ...NINKNQS  Fla  Ec
   .......... .......... .....MGFRI NTNIGALNAH A......... .......... ...NSVVNAR  FlaA Cj
   .......... .......... .....MGFRI NTNVAALNAK A......... .......... ...NADLNSK  FlaB Cj
   .......... .......... .....MAQVI NTNSLSLLTQ N......... .......... ...NLNKSQS  Fla  St

   EYNKAKKRYR DAVALVNKAG GAKKDAYLAD LQKEYETYVF KANPKSGEAR VATYIDAYNY ATKLDEMRQE
   ALSSSIERL. .......... .......... .......... ....SSG.LR INS....... ..........
   ELDKSLSRL. .......... .......... .......... ....SSG.LR INS....... ..........
   SLDASLSRL. .......... .......... .......... ....SSG.LR INS....... ..........
   ALGTAIERL. .......... .......... .......... ....SSG.LR INS....... ..........

   LEAAVQAKDL EKAEQYYHKI PYEIKTRTVI LDRVYGKTTR DLLRSTFKAK AQELRDSLIY DITVAMKARE
   .......... .......... .......... A KDDAAGQAIA NRFTSNIKGL TQAARNA.ND GISVAQTTEG
   .......... .......... .......... A ADDASGMAIA DSLRSQAATL GQAINNG.ND AIGILQTADK
   .......... .......... .......... A ADDASGMAIA DSLRSQANTL GQAISNG.ND ALGILQTADK
   .......... .......... .......... A KDDAAGQAIA NRFTANIKGL TQASRNA.ND GISIAQTTEG

   VQDAVKAGNL DKAKAAVDQI NQYLPKVTDA FKTELTEVAK KALDADEAAL TPKVESVSAI NTQNKAVELT
   ALSEINN.NL QRIRELTVQA STGT...... .......NSD SDLDSIQDEI KSRLDEIDRV SGQ.......
   AMDEQLK.IL DTIKTKATQA AQDG...... .......QSL KTRTMLQADI NRLMEELDNI ANT.......
   AMDEQLK.IL DTIKTKATQA AQDG...... .......QSL KTRTMLQADI NRLMEELDNI ANT.......
   ALNEINN.NL QRVRELAVQS ANST...... .......NSQ SDLDSIQAEI TQRLNEIDRV NGQ.......

   AVPVNGTLKL QLSAAANEDT VNVNTVRIYK VDGNIPFALN TADVSLSTDG KTITVDASTP FENNTEYKVV
   .TQFNGVNVL AKDGS.MKIQ VGAN...... .......... .......DG QTITIDLK.. ..........
   .TSFNGKQLL SGNFINQEFQ IGAS...... .......... .......SN QTVKASIG.. ..........
   .TSFNGKQLL SGNFINQEFQ IGAS...... .......... .......SN QTIKATIG.. ..........
   .TQFSGVKVL AQDNT.LTIQ VGAN...... .......... .......DG ETIDIDLK.. ..........

   VKGIKDKNGK EFKEDAFTFK LRNDAVVTQV FGTNVTNNTS VNLAAGTFDT DDTLTVVFDK LLAPETVNSS
   .......... .......... .......... .......... .......... .......... .....KIDSD
   .......... .......... .......... .......... .......... .......... .....ATQSS
   .......... .......... .......... .......... .......... .......... .....ATQSS
   .......... .......... .......... .......... .......... .......... .....QINSQ

   NVTITDVETG KRIPVIASTS GSTITITLKE ALVTGKQYKL AINNVKTLTG YNAEAYELVF TANASAPTVA
   TLGLNGFNVN GSGTIA.NKA .......... .......... .....ATISD LTAAK..... .MDAATNTIT
   KIGLTRFETG SRISVGGEVQ .......... .......... .....FTLKN YNGID..... .DFKFQKVVI
   KIGLTRFETG SRISVGGEVQ .......... .......... .....FTLKN YNGID..... .DFKFQKVVI
   TLGLDTLNVQ QKYKVS.DTA .......... .......... .....ATVTG YADTT..... .IALDNSTFK

   TAPTTLGGTT LSTGSLTTNV WGKLAGGVNE AGTYYPGLQF TTTFATKLDE STLADNFVLV EKESGTVVAS
   TTNNALTASK A......... .......... .......... ......LDQ LKDGDTVTIK ADAA..QTAT
   STSVGTGLGA L......... .......... .......... ......AEE INKSADQTGV RATF..TVET
   STSVGTGLGA L......... .......... .......... ......RDE INKNADKTGV RATF..TVET
   ASATGLGG.. .......... .......... .......... ......TDE KIDG...... ..........

   ELKYNADAKM VTLVPKADLK ENTIYQIKIK KGLKSDKG.I ELGTVNEKTY EFKTQDLTAP TVISVTSKNG
   VYTYNASAGN FSFS...... .NVSNNTSAK AGDVAASLLP PAGGTASGVY KAASGEVNFD VDANGKITIG
   RGMGAVRAGA TSED...... .FAINGVKIG QIEYKD.... ..G....... ..DANGALVS AINSVKDTTG
   RGMGAVRAGA TSDD...... .FAINGVKIG KVDYKD.... ..G....... ..DANGALVS AINSVKDTTG
   .......... .......... .......... .......... .......... .......... .DLKFDDTTG

   DAGLKVTEAQ EFTVKFSENL NTFNATTVSG STITYGQVAV VKAGANLSAL TASDIIPASV EAVTGQDGTY
   GQEAYLTSDG NLTTNDAG.. .......... ....GATAA TLDGL..... ........FK KAGDGQSIGF
   .VEASIDENG KLLLTSRE.. .......... ....GR... ..GI...... ........KI EGNIGRGAFI
   .VEASIDENG KLLLTSRE.. .......... ....GR... ..GI...... ........KI EGNIGRGAFI
   KYYAKVTVTG .......... .......... .......... .......... .......... ..GTGKDGYY

   KVKVAANQLE RNQGYKLVVF GKGATAPVKD AANANTLATN YIYTFTTEGQ DVTAPTVTKV FKGDSLKDAD
   NKTASVTMGG .......... .......... .......... TTYNFKT.GA DAGAATANAG VSF.......
   NPNMLENYGR .......... .......... .......... LSLVKND.GK DILISGTNLS AIG.......
   NPNMLENYGR .......... .......... .......... LSLVKND.GK DILISGTNLS AIG.......
   EVSVDKTNGE .......... .......... .......... VTLAAVT.PA TVTIATALSG KMY.......

   AVTTLTNVDA GQKFTIQFSE ELKTSSGSLV GGKVTVEKLT NNGWVDAGTG TTVSVAPKTD ANGKVTAAVV
   .......... .......... .......... .......... ....TDTASK ETVLNKVATA KQQTAVAANG
   .......... .......... .......... .......... ....FGTGNM ISQASVSLRE SKGQIDANVA
   .......... .......... .......... .......... ....FGTGNM ISQASVSLRE SKGQIDANVA
   .......... .......... .......... .......... ....SANA.. DSDIAKAALT AAGVTGTASV

   TLTGLDNNDK DAKLRLVVDK SSTDGIADVA GNVIKEKDIL IRYNSWRHTV ASVKAAADKD GQNASAAFPT
   DTSATITYKS ........G VQTYQAVFAA G......... .........D GTASAKYADN T.........
   DAMGFNSANK ........G NILGGYSSVS A......... .........Y MSSTGSGFSS G.........
   DAMGFNSANK ........G NILGGYSSVS A......... .........Y MSSTGSGFSS G.........
   VKMSYTDNNG ........K TIDGGLAVKV G......... .........D DYYSATQDKD G.........

   STAIDTTKSL LVEFNETDLA EVKPENIVVK DAAGNAVAGT VTALDGSTNK FVFPTPSQELK AGTVYSVTID
   .......... .......... .......... ..DVSNATAT YTDADGEMTT IGSYT..... ..........
   .......... .......... .......... ..SGFSVGSG KNYSTGFANT IAIS...... ..........
   .......... .......... .......... ..SGFSVGSG KNYSTGFANT IAIS...... ..........
   .......... .......... .......... ..SISIDTTK YTADNGTSKT ALN....... ..........
```

Fig. 4.1. Alignment of the S-layer protein SbsA with flagellins from different species. SbsA Bs, S-layer protein SbsA of B. stearothermophilus,[12] Fla Ec, Flagellin of E.coli,[54] FlaA Cj, Flagellin A, of Campylobacter jejunii,[55] FlaB, Flagellin B of Campylobacter jejunii,[55] Fla St, phase-1 i Flagellin of Salmonella typhimurium.[56] Residues shown in boldface type are identical if at least one amino acid residue of the flagellin proteins is identical with the SbsA sequence in the alignment.

was found with S-layer proteins of the remaining 20 species. The major homology regions between these S-layer proteins belong to the C-terminal and internal regions, whereas less similarity was found within the first 200 amino acids of the N-terminal regions. The little overall homology of S-layer proteins for which the corresponding genes have been cloned confirm the diversity of S-layers observed by biochemical investigations and provide further indication of the non-conservative character of these structures (see also chapters 2 and 3).

In our laboratory, using the GCG-programm BestFit[53] surprisingly revealed an overall identity (similarity) of SbsA with different flagellins (Table 4.4). As seen in Tables 4.3-4.4, nearly the same degree of identity was obtained comparing the flagellins of *Escherichia coli*,[54] *Campylobacter jejunii*,[55] and *Salmonella typhimurium*[56] with the S-layer protein SbsA as was observed with other S-layer proteins and SbsA.

The program BEAUTY (BLAST Enhanced Alignment Utility) which gives detailed information of protein families was also used for the homology studies.[52] This search revealed the most striking identity of SbsA with the protein families of flagellins, adhesins and haemagglutinins.

The multiple alignment of SbsA with different S-layer proteins showed no conserved amino acid residues at defined positions with the exception of the residue glycine (data not shown). This is in contrast to the multiple alignment of SbsA with different flagellins which revealed conserved amino acid residues (Fig. 4.1) at defined positions. The proteins displayed a conservation among the amino acids: glutamine, threonine, asparagine, arginine, serine, aspartic acid, glycine, isoleucine, alanine valine. Whether this conservation of amino acid usage depends upon codon usage alone or represents a conservation based on structure/function, e.g., protein-protein interactions, is not yet apparent.

4.5. S-LAYER PROTEIN VARIATIONS

S-layer protein variations for an individual bacterial species has been reported in four cases. This phenotypic variation of S-layer proteins is based on the existence of one or more silent S-layer genes where only one gene is expressed at a given time. The mechanisms by which such variations in S-layer protein expression occur may involve rearrangement of complete (silent) S-layer gene copies or reassembly of partial coding gene segments to form active genes.

The first silent (cryptic) S-layer gene was discovered in *Bacillus sphaericus* 2362.[11] The structural wild-type S-layer gene *gene 125* of *B. sphaericus* encodes a protein of 122 kDa, and was capable of being expressed in *E. coli, B. subtilis* and *B. sphaericus* in contrast to the cryptic S-layer gene (*gene 80*). This silent gene would encode a protein of 745 amino acids with a deduced molecular weight of 80 kDa. The 5′upstream region of *gene 80* was not preceded by a recognizable ribosome binding site or promoter region. The amino acid sequence of the putative 80 kDa protein showed extensive homology to that of the 122 kDa protein, with the exception of a portion of 127 amino acids on the C-terminus.[11]

A silent S-layer gene which is not expressed has also been found in *Lactobacillus acidophilus* (Boot HJ; personal communication). In addition to the structural gene, *slpA*,[22] which encodes an S-layer protein with a MW of 44 kDa, a silent gene (*slpB*) was also identified. Comparison of the nucleotide sequences shows the presence of two regions that are nearly identical in *slpA* and *slpB*. One region overlaps the start codon while the other overlaps the stop codon. Sequencing results from the upstream regions of *slpA* and *slpB* revealed no recognizable promoter sequence upstream of the *slpB* region. This finding is in accordance with the results that only the *slpA* gene is expressed in wild-type cells. It has been suggested that the regions of high similarity of these genes are involved in chromosomal recombination in vivo, thereby allowing the expression of *slpB*. This recombination event presumably leads to the production of an altered S-layer protein with altered properties.

The best studied model in terms of S-layer protein variation are the S-layer proteins of *Campylobacter fetus*. *C. fetus* is a Gram-negative bacterium that causes infertility and infectious abortion in sheep and cattle and extraintestinal infections in humans and expresses several variable high-molecular-weight (95-149 kDa) proteins that form S-layer lattices.[57] The S-layer protein SapA of *C. fetus* has been shown to be important for resistance to host immune defenses (see also chapter 5).[58] The wild-type strain possesses the *sapA* gene which encodes a 97 kDa S-layer protein and, in addition several *sapA* homologs. The *sapA* homologs, *sapA1* and *sapA2*, have been cloned and sequenced.[14,15] Amino acid sequence comparison of *sapA* homologs indicated two regions of identity, the first beginning 74 bp before the ORF and proceeding 552 bp into the ORF of each *sapA* homolog. The second region of identity began six nucleotides from the translation stop codon of each *sapA* homolog and extended for 51 basepairs. The presence of these homologous sequences at both the 5′ and 3′ ends of these *sapA* homologs suggested that these regions might play a role in homologous recombination. As sequence analysis did not demonstrate typical promoter sequences 5′upstream of *sapA1* and *sapA2*, it was suggested that variations of *C. fetus* S-layer proteins involve rearrangement of silent gene cassettes into a unique expression locus.[15] The promoter upstream of *sapA* is the only such locus that has been identified, and loss of this promoter in spontaneous mutants is associated with loss of transcription and S-layer protein expression.[37] Although the *sapA* homologs undergo recombinatorial events on the order of 10^{-2} to 10^{-3} per generation these genes are remarkably stable. A stable genotype in spite of frequent recombinatorial events could be due to a mechanism involving duplicative transfer of genetic information from a silent locus to the expression site, thereby preserving the original silent genetic elements.[15] The stability of the coding region also suggests conservation of protein function, which has been clearly demonstrated for the N-terminal region. Although spontaneous S-layer variants of wild-type *C. fetus* strains have been observed, the induction of S-layer protein variation is mediated by incubation with serum and mouse passage where selection goes towards strains which express S-layer proteins because the proteins protect bacteria against complement mediated killing.

In the case of *Bacillus stearothermophilus* PV72, the change in S-layer protein synthesis between the SbsA and SbsB type is correlated with physiological changes in cultures.[59] *B. stearothermophilus* PV72 is a strictly aerobic organism and under oxygen-limited conditions the S-layer protein gene *sbsA* is expressed which codes for a protein with a MW of 130-kDa (hexagonal lattice, p6; see

Table 4.1). When oxygen limitation is relieved, the S-layer protein SbsA becomes rapidly replaced by a new type of S-layer protein SbsB with a MW of 97 kDa which assembles into an oblique lattice (p2). This irreversible change of S-layer proteins has been observed in a number of *B. stearothermophilus* strains.[60] In the case of *B. stearothermophilus* PV72/p2, the N-terminal (mature protein) and internal amino acid sequences are not identical with the S-layer gene *sbsA* encoding the p6-S-layer protein, suggesting that the two S-layer proteins are encoded by different genes. The N-terminal region of the *sbsB* gene has been cloned and sequenced, and the deduced amino acid sequences of *sbsA* and *sbsB* have been aligned and indicate a similarity of 53% over a stretch of 112 amino acids.[61] At present there are no data available as to whether the *sbsB* gene retains major regions of identity with the *sbsA* gene or whether a promoter region upstream of *sbsB* is present or missing.

4.6. S-LAYER DOMAINS

This section deals with S-layer domains as determined by biochemical and molecular-biological methods (see also chapter 3). A domain is defined as a polypeptide chain or a part of a polypeptide chain that can independently fold into a stable tertiary structure. Domains are also units of function, with the different domains of a protein often associated with different functions.[62]

4.6.1. TWO MAJOR STRUCTURAL DOMAINS OF THE S-LAYER PROTEIN OF *A. SALMONICIDA*

The S-layer of *Aeromonas salmonicida* (previously termed A-layer) is composed of a single protein species with 481 amino acid residues and a molecular weight of 50,8 kDa.[6] Image analysis of S-layer lattices have shown that the constituent subunits contain a heavy mass domain with a linker arm to a domain of lesser mass.[63] The major tetragonal core of the array is composed of the heavy mass domains of four subunits, contains a large depression, and is oriented towards the inside of the latter.[63] Protease digestion studies with purified S-proteins have provided biochemical evidence for two major structural domains. Treatment with trypsin produced an N-terminal peptide of an apparent MW of 35.5 kDa by SDS-PAGE which was totally refractile to tryptic digestion despite the presence of 27 trypsin cleavage sites, and an overlapping 16.7 kDa C-terminal peptide. This suggested that the two structural segments reflected the two morphological domains of the *Aeromonas* S-layer proteins, the C-terminal peptide comprising the lesser morphological linker domain, and the N-terminal peptide corresponding to the inner major mass morphological core domain. Immunological methods have been used to map the antigenic structure of the S-layer protein.[64] The results indicated that the major N-terminal domain carries the bulk of non-surface exposed sequences, consistent with the region of S-layer protein which interacts with the heavy mass domains to form the major core of the surface array which is located towards the cell surface side of the layer. The C-terminal domain carries the majority of surface exposed sequences which is consistent with its role in forming the minor array composed of the lesser mass domains. This domain provides connectivity within the layer, and is located at the outer surface of the layer.

Consistent with these interpretations for conservation of structure/function in S-layer proteins of *Aeromonas*, protease mapping studies of *Aeromonas hydrophila* TF7 has also revealed two structural domains.[65, 66]

4.6.2. DOMAIN STRUCTURE OF THE *ACETOGENUM KIVUI* S-LAYER PROTEIN

To define the domains of the *A. kivui* S-layer protein several methods were used. Three-dimensional reconstructions revealed three domains for the monomer, the C-, Y- and L-domain. Further, a domain from residue 27 to approximately residue 200 (18 kDa) was defined by sequence homology to the outer wall of *B. brevis* (SLH-domain; see 4.6.6.). Proteolysis with proteinase K yields two further protease resistant domains, one from residue 215 to approximately 475 (33 kDa) and one from residue 480 to approximately residue 634 (15 kDa). The boundaries of these domains correspond closely to those of the α (residues 27 to 217), β (residues 218 to 480), and mixed α and β (residues 481 to 736) domains defined by secondary-structure prediction. It was suggested that the N-terminal SLH region forms the bottom of the core, the protease resistant 33 kDa domain forms the upper part of the core and the short arm of the Y-domain, and the protease resistant

15 kDa domain forms the long arm of the Y-domain. This arrangement was most compatible with the location of the glycosylation sites, which then lie in the long arm of the Y-domain and thus in the most exposed part of the layer.[67]

4.6.3. LPS-BINDING DOMAIN

Campylobacter fetus cells can produce multiple S-layer proteins ranging in size from 97 to 149 kDa. Cloning and subsequent sequence analysis of the genes *sapA*, *sapA*1 and *sapA*2 has demonstrated that the S-layer proteins share identical N-termini, extending 219 amino acid residues into the proteins but which diverge for the remainder of the molecule.

Reattachment assays with isolated S-layer subunits indicated that S-layer proteins of *C. fetus* exclusively adhere to the LPS layer at the outer membrane in a serotype (A, B) specific manner. Deletion mutagenesis and reattachment assays of truncated S-layer proteins revealed that the S-layer protein bound sero-specifically to the LPS via its conserved N-terminal region. These data indicated that S-layer proteins shared functional activity in the conserved N-terminus.[15]

4.6.4. CALCIUM-BINDING DOMAIN

Investigations on the manner of attachment of the *Caulobacter crescentus* S-layer protein RsaA to the cell surface indicated that calcium somehow affects S-layer attachment and self-assembly.[68] Mutants of *Caulobacter crecentus,* isolated for their ability to grow in the absence of calcium ions, no longer had the S-layer attached to the cell surface and failed to produce a cell surface polysaccharide molecule. This molecule was shown to be a smooth LPS with a sugar core (SLPS).[68] Thus it was proposed that calcium is involved in binding to the SLPS and, in turn, to the assembled RsaA protein. Moreover DNA sequence analysis of *rsaA* demonstrates regions of homology with calcium-interactive regions of other calcium required bacterial proteins.[16]

4.6.5. C-TERMINAL HYDROPHOBIC MEMBRANE BINDING DOMAIN

S-layer proteins of *Corynebacterium glutamicum,*[17] *Halobacterium halobium,*[19] *Halobacterium volcanii*[20] and *Rickettsia prowazekii*[25] exhibit a hydrophobic domain of 21 residues near the C-terminus proposed to be a membrane-binding domain.

4.6.6. SLH-DOMAIN (S-LAYER HOMOLOGY)
FOR PEPTIDOGLYCAN ANCHORAGE

The SLH-domain was obtained by sequence alignment of several S-layer proteins.[67] The SLH sequences are strongly divergent, with an average identity of only 27%, although a single residue, Gly-28, is universally conserved.[67] SLH sequences have been found either at the very beginning or at the very end of proteins. This domain was found in the S-layer proteins of *Acetogenum kivui,*[4] *Bacillus brevis,*[8,10] *Bacillus sphaericus,*[11] *Thermus thermophilus*[28] and in several other extracellular proteins, including the Ompα from *Thermotoga maritima,*[69] an alkaline cellulase from Bacillus strain KSM-635,[70] a xylanase from *Clostridium thermocellum,*[71] and a pullulanase of *Thermoanaerobacterium thermosulfurigenes* EM1.[72] It has been suggested that SLH-sequences serve as an anchor to the peptidoglycan (see also chapter 5).[67]

4.7. RECOMBINANT S-LAYER GENES AND MOLECULAR BIOTECHNOLOGY

Since S-layers represent one of the most commonly observed bacterial surface structures and several sequences have been elucidated, a great potential for using S-layers exists. The possibility to build two-dimensional crystalline arrays of identical protein or glycoprotein subunits on different surfaces or interfaces opens appealing possibilities for functioning surfaces and to build supramolecular structures in the third dimension (see chapter 8). Most importantly, S-layer lattices are highly anisotropic structures exhibiting remarkable differences in the topography and physicochemical properties of their surfaces (see chapter 2; Fig. 2.1 and chapter 5). In nature, the inner surface often interacts either with peptidoglycan or membrane surfaces.

So far the distribution of hydrophobic, hydrophilic and charged residues on the surface of the protomers have only been determined by using charged marker molecules (see chapters 6 and 8). Until more detailed information on the arrangement of single amino acids or atoms within the tertiary structure of the S-layers

is available, the recombinant modification by exchange or introduction of single amino acids, modulation of epitopes within
surface loops, weakening or enforcing intra-and intermolecular interactions will remain the most practical tools to functionally
describe the molecular architecture of specific S-layers.

One approach is to localize surface areas which allow the insertion (or loss) of amino acids without major changes in the
structural properties of the mono- and protomers and the capability for assembly into two-dimensional protein lattices. An easy way
to achieve multiple inserts with a simple selection procedure would
be insertion mutagenesis with a modified antibiotic resistance cassette. Preliminary studies on the S-layer protein of *Caulobacter
crescentus* identified potential sites for foreign epitope insertion,
using a modified random linker mutagenesis method. Such sites
can tolerate extra amino acids without loss of biological properties
and therefore without major changes in cellular location and
structure. [73,74]

Another approach which was used for insertion mutagenesis in
the case of the S-Layer protein gene *sbsA* of *B. stearothermophilus
PV72* is described here in detail.[75] Partial digestion of cloned S-layer
sequences using an enzyme with multiple sites within the S-layer
gene sequence allows insertion of the cassette at several positions
within the gene in a single procedure. A selection based on antibiotic resistance is carried out and only mutagenized clones are
able to grow. Plasmid DNA is isolated from the single clones and
the cassette removed by excision leaving behind an insert with three
nucleotides preferentially coding for a rare restriction site. In the
case of *Apa*I, six bp (GGGCCC) are inserted and result in an
extension of two amino acids at the place of mutation coding in
frame for glycine and proline. In the case of heterologous expression of the *B. stearothermophilus* PV72 S-layer gene *sbsA* in *E. coli*,
self-assembly of the S-layer protein is indicated by sheet formation
within the cells (Kuen B, Sára M, Lubitz W. Heterologous expression and self-assembly of the S-layer protein SbsA of *Bacillus
stearothermophilus* in *Escherichia coli*. Submitted 1995). In this
manner electron microscopic examinations indicate whether the
different inserts interfere with inter- and intramolecular interactions involved in self-assembly process. When the extension of such

tolerated insertion sites is made by sequences coding for epitopes which can be recognized by specific antibodies or by sequences coding for protease cleavage sites, the surface location of such extensions are then accessible for experimental test procedures. Insertion of streptavidin, for example, within a surface loop provides the possibility to attach any biotinylated substance at a defined position within the topology of S-layer proteins. Integration of protein A or protein G would bind the Fc-part of antibodies and the variable side chains of the antibodies could bind other specific proteins. These few examples show that the architectural principles using S-layer structures as rigid anchors can be applied to modulate the surfaces and/or to build up multiple, multifunctional sandwich layers with various functions (see also chapter 8).

ACKNOWLEDGMENT

We thank Karsten Tedin for critical reading the manuscript. This work was supported by the Austrian Science Foundation, Project SO7208.

REFERENCES

1. Baumeister W, Engelhardt H. Three-dimensional structure of bacterial surface layers. In: Harris JR, Horne RW, eds. Electron microscopy of proteins. vol 6. London Academic Press, 1987:109-54.
2. Beverdige TJ, Koval SF, eds. Advances in paracrystalline surface layers. New York: Plenum Press, 1993.
3. Sleytr UB, Messner P, Pum D et al. Crystalline bacterial cell surface layers. Mol Microbiol 1993; 10:911-16.
4. Peters J, Peters M, Lottspeich F et al. S-layer protein gene of *Acetogenium kivui:* cloning and expression in *Escherichia coli* and determination of the nucleotide sequence. J Bacteriol 1989; 171:6307-15.
5. Thomas SR, Trust TJ. Tyrosine phosphorylation of the tetragonal paracrystalline array of *Aeromonas hydrophila:* molecular cloning and high-level expression of the S-layer protein gene. J Mol Biol 1995; 245:568-81.
6. Chu S, Cavaignac S, Feutrier J et al. Structure of the tetragonal surface virulence array protein and gene of *Aeromonas salmonicida.* J Biol Chem 1991; 266:15258-65.
7. Etienne-Toumelin I, Sirard JC, Duflot E et al. Characterization of the *Bacillus anthracis* S-layer: cloning and sequencing of the structural gene. J Bacteriol 1995; 177:614-20.

8. Tsuboi A, Uchihi R, Adachi T et al. Characterization of the genes for the hexagonally arranged sufrace layer proteins in protein-producing *Bacillus brevis* 47: complete nucleotide sequence of the middle wall protein gene. J Bacteriol 1988; 170:935-45.

9. Tsuboi A, Uchihi R, Tabata R et al. Characterization of the genes coding for two major cell wall proteins from protein-producing *Bacillus brevis* 47: complete nucleotide sequence of the outer wall protein gene. J Bacteriol 1986; 168:365-73.

10. Ebisu S, Tsuboi A, Takagi H et al. Conserved structures of cell wall protein genes among protein-producing *Bacillus brevis* strains. J Bacteriol 1990; 172:1312-20.

11. Bodwitch RD, Baumann P, Yousten AA. Cloning and sequencing of the gene encoding a 125-kilodalton surface-layer protein from *Bacillus sphaericus* 2362 and of a related cryptic gene. J Bacteriol 1989; 171:4178-88.

12. Kuen B, Sleytr UB, Lubitz W. Sequence analysis of the *sbsA* gene encoding the 130-kDa surface-layer protein of *Bacillus stearothermophilus* strain PV72. Gene 1994; 145:115-20.

13. Blaser MJ, Gotschlich EC. Surface array protein of *Campylobacter fetus*. J Biol Chem 1990; 265:14529-35.

14. Tummuru MK, Blaser MJ. Rearrangements of *sapA* homologs with conserved and variable regions in *Campylobacter fetus*. Proc Natl Acad Sci USA 1993; 90:7265-69.

15. Dworkin J, Tummuru MKR, Blaser MJ. A lipopolysaccharide-binding domain of the *Campylobacter fetus* S-layer protein resides within the conserved N-Terminus of a family of silent and divergent homologs. J Bacteriol 1995; 177:1734-41.

16. Gilchrist A, Fisher JA, Smit J. Nucleotide sequence analysis of the gene encoding the *Caulobacter crescentus* paracrystalline surface layer protein. Can J Microbiol 1992; 38:193-202.

17. Peyret JL, Bayan N, Joliff G et al. Characterization of the *cspB* gene encoding PS2, an ordered surface-layer protein in *Corynebacterium glutamicum*. Mol Microbiol 1993; 9:97-109.

18. Peters J, Peters M, Lottspeich F et al. Nucleotide sequence analysis of the gene encoding the *Deinococcus radiodurans* surface protein, derived amino acid sequence, and complementary protein chemical studies. J Bacteriol 1987; 169:5216-23.

19. Lechner J, Sumper M. The primary structure of a procaryotic glycoprotein. J Biol Chem 1987; 262:9724-29

20. Sumper M, Berg E, Mengele R et al. Primary structure and glycosylation of the S-layer protein of *Haloferax volcanii*. J Bacteriol 1990; 172:7111-18.

21. Vidgren G, Palva I, Pakkanen R et al. S-layer protein gene of *Lactobacillus brevis*: cloning by polymerase chain reaction and determination of the nucleotide sequence. J Bacteriol 1992; 174:7419-27.

22. Boot JH, Kolen CPAM, van Noort JM et al. S-layer protein of *Lactobacillus acidophilus* ATCC 4356: purification, expression in *Escherichia coli*, and nucleotide sequence of the corresponding gene. J Bacteriol 1993; 175:6089-96.

23. Bröckl G, Behr M, Fabry S et al. Analysis and nucleotide sequence of the genes encoding the surface-layer glycoproteins of the hyperthermophilic methanogens *Methanothermus fervidus* and *Methanothermus sociabilis*. Eur J Biochem 1991; 199:147-52.

24. Dharmavaram R, Gillevet P, Konisky J. Nucleotide sequence of the gene encoding the vanadate-sensitive membrane-associated ATPase of *Methanococcus voltae*. J Bacteriol 1991; 173:2131-33.

25. Carl M, Dobson ME, Ching WM et al. Characterization of the gene encoding the protective paracrystalline-surface-layer protein of *Rickettsia prowazekii*: presence of a truncated identical homolog in *Rickettsia typhii*. Proc Natl Acad Sci USA 1990, 87:8237-41.

26. Gilmore RD, Joste N, McDonald GA. Cloning, expression and sequence analysis of the gene encoding the 120 kD surface-exposed protein of *Rickettsia rickettsii*. Mol Microbiol 1989; 3:1579-86.

27. Hahn MJ, Kim KK, Kim I et al. Cloning and sequence analysis of the gene encoding the crystalline surface layer protein of *Rickettsia typhii*. Gene 1993; 133:129-33.

28. Faraldo MM, de Pedro MA, Berenguer J. Sequence of the S-layer gene of *Thermus thermophilus* HB8 and functionality of its promoter in *Escherichia coli*. J Bacteriol 1992; 174:7458-62.

29. Messner P, Sleytr UB. Crystalline bacterial cell-surface layers. Adv Microb Physiol 1992; 33:213-75.

30. Adachi T, Yamagata H, Tsukagoshi N et al. Multiple and tandemly arranged promoters of the cell wall protein gene operon in *Bacillus brevis* 47. J Bacteriol 1989; 171:1010-16.

31. Chu S, Gustafson CE, Feutrier J et al. Transcriptional analysis of the *Aeromonas salmonicida* S-layer protein gene *vapA*. J Bacteriol 1993; 175:7968-75.

32. Chu S, Trust TJ. An *Aeromonas salmonicida* gene which influences A-protein expression in *Escherichia coli* encodes a protein containing an ATP-binding cassette and maps beside the Surface Array protein gene. J Bacteriol 1993; 175:3105-14.

33. Chu S, Noonan B, Cavaignac S et al. Endogenous mutagenesis by an insertion sequence element identifies *Aeromonas salmonicida* AbcA as an ATP-binding cassette transport protein required for biogenesis of smooth lipopolysaccharide. PNAS 1995; 92:5754-58.

34. Noonan B, Trust TJ. The leucine zipper of *Aeromonas salmonicida* AbcA is required for the transcriptional activation of P2 promoter of the surface-layer structural gene, *vapA*, in *Escherichia coli*. Mol Microbiol 1995; 379-86.

35. Kansy JW, Carinato ME, Monteggia LM et al. In vivo transcripts of the S-layer-encoding structural gene of the archaeon *Methanococcus voltae*. Gene 1994; 148:131-135.

36. Thomm M, Wich G, Brown JW et al. An archaebacterial promoter sequence assigned by RNA polymerase binding experiments. Can J Microbiol 1989; 35:30-35.

37. Tummuru MKR, Blaser MJ. Characterization of the *Campylobacter fetus sapA* promoter: evidence that the *sapA* promoter is deleted in spontaneous mutant strains. J Bacteriol 1992; 174:5916-22.

38. Hawley DK, McClure WR. Compilation and analysis of *Escherichia coli* promoter DNA sequences. Nucleic Acids Res 1983; 11:2237-55.

39. Fisher JA, Smit J, Agabain N. Transcriptional analysis of the major surface array gene of *Caulobacter crescentus*. J Bacteriol 1988; 170:4706-13.

40. Grundy FJ, Henkin, TM. Cloning and analysis of the *Bacillus subtilis rpsD* gene, encoding ribosomal protein S4. J Bacteriol 1990; 172: 6372-79.

41. Nilsson, G, Belasco JG, Cohen SN et al. Growth rate dependent regulation of mRNA stability in *Escherichia coli*. Nature 1984; 312:75-77.

42. Belasco JG, Nilsson G, von Gabain A et al. The stability of *E. coli* gene transcripts is dependent on determinants localized to specific mRNA segments. Cell 1986; 46:245-51.

43. Emory SA, Bouvet P, Belasco JG. A 5′terminal stem-loop structure can stabilize mRNA in *Escherichia coli*. Genes Dev 1992; 6:135-48.

44. Causton H, Py B, Mc Laren RS. mRNA degradation of *E. coli*: a novel factor which impedes the exoribonucleolytic activity of PNPase at stem-loop structures. Mol Microbiol 1994; 14: 731-41.

45. Nelson DR, Zusman DR. Evidence for long-lived mRNA during fruiting body formation in *Myxococcus xynthus*. Proc Natl Acad Sci USA 1983; 80:12589-95.

46. Romeo JM, Zusman DR. Determinants of an unusually stable mRNA in the bacterium *Myxococcus xanthus*. Mol Microbiol 1992; 6:2875-88.

47. Belland RJ, Trust TJ. Cloning of the gene for the surface array protein of *Aeromonas salmonicida* and evidence linking loss of expression with genetic deletion. J Bacteriol 1987; 169:4086-91.

48. Miymoto M, Kobayashi Y, Kokeguchi S et al. Molecular cloning of the S-layer protein gene of *Campylobacter rectus* ATCC 33238. FEMS Microbiol Lett 1994; 116:13-18.

49. Bingle WH, Smit J. Alkaline phosphatase and a cellulase reporter protein are not exported from the cytoplasm when fused to large N-terminal portions of the *Caulobacter crecentus* surface (S)-layer protein. Can J Microbiol 1994; 777-82.

50. Noonan B, Trust TJ. Molecular analysis of an A-protein secretion mutant of *Aeromonas salmonicida* reveals a surface layer -specific protein secretion pathway. J Mol Biol 1995; 248:316-27.

51. Thomas SR, Trust TJ. A specific PulD homolog is required for the secretion of paracrystalline surface array subunits in *Aeromonas hydrophila*. J Bacteriol 1995; 177:3932-39.

52. Altschul SF, Gish W, Miller W et al. Basic local alignment search tool. J Mol Biol 1990; 215:403-10.

53. Bestfit-programm: Genetics computer group (University of Wisconsin) package.

54. Kuwajina G, Asaka JI, Fujiwara T, et al. Nucleotide sequence of the hag gene encoding flagellin of *Escherichia coli*. J Bacteriol 1986; 168: 1479-1483.

55. Nuijten PJM, van Asten FJAM, Gaastra W et al. Structural and functional analysis of two *Campylobacter jejuni* flagellin genes. J Biol Chem 1990; 265:17798-804.

56. Joys TM. The covalent structure of the phase-1 flagellar filament protein of *Salmonella typhimurium* and its comparison with other flagellins. J Biol Chem 1985; 260:15758-61.

57. Dubreuil JD, Kostrzynska M, Austin JW et al. Antigenic differences among *Campylobacter fetus* S-layer proteins. J Bacteriol 1990; 172:5035-43.

58. Blaser MJ, Wang E, Tummuru MKR et al. High-frequency S-layer protein variation in *Campylobacter fetus* revealed by *sapA* mutagenesis. Mol Microbiol 1994; 14:453-62.

59. Sára M, Sleytr UB. Comparative studies of S-layer proteins from *Bacillus stearothermophilus* strains expressed during growth in continuous culture under oxygen-limited and non-oxygen-limited conditions. J Bacteriol 1994; 176:7182-89.

60. Sára M, Pum D, Küpcü S et al. Isolation of two physiologically induced variant strains of *Bacillus stearothermophilus* NRS 2004/3a and characterization of their S- layer lattices. J Bacteriol 1994; 176:848-60.

61. Asenbauer E. Klonierung und Charakterisierung der 5′seitigen Sequenz des S-layer Gens *sbsB* von *Bacillus stearothermophilus* PV72. Master thesis. Universität Wien.1995.

62. Brandon C, Tooze J, eds. Introduction to protein structure. New York: Garland 1991.

63. Dooley JSG, McCubbin WD, Kay CM et al. Isolation and biochemical characterization of the S-layer protein from a pathogenic strain of *Aeromonas hydrophila*. J Bacteriol 1988; 170:2631-38.

64. Doig P, McCubbin WD, Kay CM et al. Distribution of surface-exposed and non-accessible amino acid sequences among the two major structural domains of the S-layer protein of *Aeromonas salmonicida*. J Mol Biol 1993; 233:753-65.

65. Kostrzynska M, Dooley JSG, Shimojo T et al. Antigenic diversity of the S-layer proteins from pathogenic strains of *Aeromonas hydrophila* and *Aeromonas veronii* biotype sobria. J Bacteriol 1992; 174:40-7.

66. Austin TS, McCubbin JW, Kay W et al. Roles of structural domains in the morphology and surface anchoring of the tetragonal paracrystalline array of *Aeromonas hydrophila*. Biochemical characterization of the major structural domain. J Mol Biol 1992; 228:652-61.

67. Lupas A, Engelhardt H, Peters J et al. Domain structure of the *Acetogenium kivui* surface layer revealed by electron crystallography and sequence analysis. J Bacteriol 1994; 176:1224-33.

68. Walker SG, Karunaratne DN, Ravenscroft N et al. Characterization of mutants of *Caulobacter crescentus* defective in surface attachment of the paracrystalline surface layer. J Bacteriol 1994; 176:6312-23.

69. Engel AM, Cejka Z, Lupas A et al. Isolation and cloning of Ompα, a coiled-coil protein spanning the periplasmic space of the ancestral eubacterium *Thermotoga maritima*. EMBO J 1992; 11:4369-78.

70. Ozaki K, Shikata S, Kawai S et al. Molecular cloning and nucleotide sequence of a gene for alkaline cellulase from *Bacillus* sp. KSM-635. J Gen Microbiol 1990; 136:1327-34.

71. Fujino T, Beguin P, Aubert JP. Organization of a *Clostridium thermocellum* gene cluster encoding the cellulosomal scaffolding protein CipA and a protein possibly involved in attachment of the cellolosome to the cell surface. J Bacteriol 1993; 175:1891-99.

72. Matuschek M, Burchhardt G, Sahm K et al. Pullulanase of *Thermoanaerobacterium thermosulfurigenes* EM1 (*Clostridium thermosulfurogenes*): Molecular analysis of the gene, composite structure of the enzyme, and a common model for its attachment to the cell surface. J Bacteriol 1994; 176:3295-02.

73. Bingle WH, Smit J. Linker mutagenesis using a selectable marker: A method for tagging specific-purpose linkers with an antibiotic resistance gene. BioTechniques 1991; 10:150-52.

74. Bingle HB, Walker SG, Smit J. Definition of Form and Function for the S-layer of *Caulobacter crescentus*. Advances in paracrystalline surface layers. New York. Plenum 1993.

75. Lechleitner S. Insertion mutagenesis in the S-layer gene *sbsA* of *Bacillus stearothermophilus* PV72. Master thesis. Universität Wien. 1995.

===== CHAPTER 5 =====

FUNCTIONAL ASPECTS OF S-LAYERS

Margit Sára, Eva Maria Egelseer

5.1. INTRODUCTION

In the course of evolution, prokaryotic organisms have developed a considerable diversity in the supramolecular architecture of their multilayered cell boundaries. One of the most commonly observed bacterial cell surface structures are two-dimensional arrays of proteinaceous subunits termed S-layers. S-layers represent an almost universal feature of the archaeobacterial cell envelopes and were already detected in hundreds of different species of nearly every taxonomic group of walled eubacteria.[1-15] S-layers represent an important class of secreted proteins and can comprise up to 15% of the cellular protein. In eubacteria, a comparable number of S-layers with oblique, square or hexagonal lattice symmetry has been identified, whereas S-layers from most archaeaobacteria exhibit hexagonal symmetry. Only a few exceptions such as the square and poorly ordered S-layer lattice from *Desulfurococcus mobilis*[16] and the sheaths from *Methanospirillum*[17-20] showing oblique lattice symmetry are known. Many archaeobacteria and some eubacteria have the capacity for glycosylating their S-layer proteins.[21]

Although a considerable knowledge has accumulated on the structure, assembly, chemistry and more recently on the genetics of S-layers, only little information is available about the specific

Crystalline Bacterial Cell Surface Proteins, edited by Uwe B. Sleytr, Paul Messner, Dietmar Pum, Margit Sára. © 1996 R.G. Landes Company.

biological function of these crystalline arrays. It has repeatedly been suggested that because S-layers can be detected on bacteria from quite different ecologies and habitats, they must fulfill a broad spectrum of functions, particularly in the hostile and competitive natural environments.

In the present article more recently described specific functions of S-layers from pathogens, cyanobacteria and exoenzyme producing Bacillaceae are summarized. Furthermore, a survey on general functional aspects such as the surface and molecular-sieving properties of these two-dimensional protein crystals is given.

5.2. SPECIFIC FUNCTIONS

5.2.1. PATHOGENICITY AND ADHESION

The S-layer of *Aeromonas salmonicida,* previously described as A layer, has been shown to be the major virulence factor responsible for pathogenicity in fish. It is composed of a single protein species with a molecular weight of approximately 50,000 (A protein), organized into a two-dimensional crystal with square symmetry.[22] The structural gene for the A protein, *vapA,* has been cloned, sequenced, expressed and the large mass and linker domains identified.[23,24] Early studies have shown that the presence of an S-layer endows the *A. salmonicida* cells with high or intermediate resistance to the bactericidal activity of complement in both immune and non-immune serum.[25] An intact S-layer was found to mediate the adherence of *A. salmonicida* to macrophages even in the absence of opsonins due to the high content of hydrophobic amino acids as well as the physical masking of the hydrophilic polysaccharide O-chains.[26] In contrast, unopsonized cells of an S-layer defective mutant with a smooth LPS-layer were unable to interact with macrophages but this ability was recovered when the S-layer was reconstituted onto the outer membrane.[27] The highest level of macrophage, association in the absence of opsonins was demonstrated with S-layer carrying cells cultivated under Ca^{2+}-limited conditions[27] under which the S-layer adopts an alternative conformation termed the "big squares" (BS).[28] Recent studies have demonstrated that the presence of an S-layer on *A. salmonicida* predisposes this bacterium to apparently unrelated physiological consequences: inhibition of growth at 30°C, enhanced cell filamentation at 37°C, and enhanced uptake of the hydrophobic antibiotics streptonigrin and chloramphenicol.[29] Disorganization of the S-layer, genetically encoded faulty assembly, lack of LPS receptors or lack of an in vivo assembly process affected all the three physiological parameters suggesting that they are coupled to some kind of interaction between a specifically ordered S-layer and the outer membrane. This could lead to hydrophobic channels involved in the uptake of the hydrophobic antibiotics.[29] As previously reported, the S-layer may also be responsible for the uptake of porphyrins.[30] The fact that in S+-cells the S-layer bound hemin was an effective antagonist of streptonigrin toxicity strongly suggests that streptonigrin and porphyrin may share at least for the first step, the same uptake mechanism.[29]

A. salmonicida cells further possess unique binding capabilities for immunoglobulins, whereas a specific arrangement of adjacent A-protein subunits such as is present in the square unit cell of the array was required for binding.[31] Recent studies have shown that the S-layer is capable of binding a variety of extracellular matrix proteins such as collagen type IV,[32] laminin and fibronectin.[33] In contrast to the S-layer of *A. salmonicida,* the biological functions of the S-layers from the mesophilic *A. hydrophila* and *A. sobria* remain obscure. The square S-layers produced by these motile aeromonads are morphologically very similar[34,35] but appear to be genetically unrelated[23,36] and lack the binding abilities of *A. salmonicida* cells.[31]

As described for *A. salmonicida,* the S-layer from *Campylobacter fetus, subsp. fetus* plays an important role in pathogenicity. This organism causes infertility and infectious abortus in sheep and cattle and was found to lead to extraintestinal infections in humans, too.[37] In *C. fetus,* the S-layer is involved in infection and is important for survival within the host: S-layer carrying strains usually resist phagocytosis by polymorphonuclear leucocytes but this anti-phagocytic ability is lost in the presence of opsonizing antibodies.[38,39] Serum-resistance of S-layer carrying strains was associated with the inability of complement component C_{3b} to bind to the cell surface. Removal of the S-layer protein with proteases had little effect on the viability of the cells, but such treated cells became susceptible to serum-mediated killing as well as phagocytosis. Two serotypes of *C. fetus* strains, based on LPS types (A and B) which function as receptors for the S-layer proteins are known.[40] Accordingly, the S-layer proteins are divided into two groups (A and B). Since type A S-layer proteins can only bind to type A LPS, and type B S-layer proteins only recognize type B LPS structures, a specific recognition mechanism must exist in which Ca^{2+}-ions are involved.[41] Regardless the molecular weight of the subunits, within one protein family, all type A and type B S-layer proteins showed identical N-terminal regions.[42]

S-layer protein synthesis in *C. fetus* is subject to antigenic variation which represents a mechanism for creating a heterogeneous population to survive environmental fluctuations.[43-45] With three clinical isolates, the molecular weight of the major S-layer protein was either 98,000, 127,000 or 149,000.[46] The S-layer protein with molecular weight 98,000 assembled into a hexagonally ordered lattice, whereas the higher molecular weight proteins formed square S-layer lattices.[47] Although this diversity in molecular weight and structure was originally interpreted to represent a strain specific feature, switching from one to another S-layer during in vivo and in vitro passage and the existence of two different types of S-layer lattices as a monomolecular layer on the cell surface indicated that changes in S-layer protein synthesis must occur within single cells.[47] Switching from one to another S-layer protein was only observed within the three molecular weights showing that the repertoire for changing the S-layer characteristics is highly limited. Immuno-blotting studies using polyclonal antibodies revealed that the S-layer proteins are antigenically related. However, the use of different monoclonal antibodies confirmed the existence of a great diversity of epitopes; some of them were specific to the molecular size of the S-layer proteins and some of them were typical of the protein family.[45]

The genus *Rickettsia* comprises obligate intracellular bacteria which cause a diverse group of human diseases as well as non pathogenic forms found only in arthropods.[48] The genus is usually divided into the typhus, spotted fever, and scrub thyphus groups. The typhus group includes *R. prowazekii* and *R. typhi*, the agents responsible for epidemic typhus and endemic typhus, respectively.

The serotype protein antigen (SPA) that comprises a microcapsule crystalline S-layer protein revealed to be immunodominant for both humoral and cell-mediated immunity of rickettsia.[49] The 120,000 and 150,000 S-layer proteins of typhus and spotted fever rickettsiae carry species-specific epitopes, are heat-labile, and elicit toxin neutralizing antibodies.[50,51] Several cellular mechanisms are stimulated by the SPA of rickettsiae: (i) T-helper cells stimulate the generation of natural killer cells and non-HLA restricted lymphokine-activated killer cells which are effective in lysing rickettsiae-infected cells.[52,53] (ii) T-helper cells regulate the specific anti-SPA antibody synthesis by B cells; these antibodies mediate opsonization and destruction of extracellular rickettsiae and (iii) T-helper cells are stimulated by the SPA to secrete γ-interferon.[49]

Clostridium difficile is an important pathogen that causes human pseudomembraneous colitis. The square S-layer lattice of this organism was found to be composed of two different types of subunits with molecular weights of either 45,000 and 32,000[54] which showed different N-terminal regions and no crossreaction with polyclonal antibodies. Immunogold-labeling of whole cells confirmed that both proteins are arranged on the cell surface as a monomolecular layer.[55] The S-layer from *C. difficile* was found to participate in adhesion of whole cells to human embryonic intestinal cells, in adult colon cells and is responsible for attachment to HeLa cells. F_{ab} fragments prepared against the 45,000 and 32,000 S-layer proteins significantly inhibited bacterial attachment confirming that both types of S-layer proteins are involved in the adhesion process.

Lactobacilli are of great importance in milk technology and represent valuable inhabitants of the intestinal and urogenital tract of humans. The molecular weight S-layer proteins from *Lactobacilli* is in the range of 40,000 to 60,000.[56-60] Generally, they possess an acid-tolerant S-layer lattice which has most likely evolved as cell surface structure quite resistant to the low pH conditions characteristic of their natural environment. For *L. acidophilus* the S-layer was found to mediate adhesion of the bacteria to epithelial cells.[61] Subcultivation of this organism on agar plates led to a markedly reduced ability to adhere to epithelial cells which correlated well with the loss of the S-layer lattice and changes in physiology. A similar observation that the frequency of appearance of an S-layer diminished and that it was most likely lost by sequentially subculturing was done for the insect pathogen *Bacillus thuringiensis* for which the S-layer was suggested to play an important role in the infection route.[62] Late-exponential and stationary growth phase cells typically sloughed off S-layer fragments which was interpreted to result from wall turnover. Intrinsic autolytic activity in isolated cell walls rapidly digested the wall fabric, liberating soluble S-layer protein. Thus it was concluded that the bonds between the subunits are weaker than those between the subunits and the underlying cell envelope layer.

B. anthracis, the causative agent of anthrax, is closely related to the entomopathogen *B. thuringiensis*. Fully virulent cells are both capsulated and toxinogenic. When the capsule is absent, the cell surface displays a regularly structured S-layer. The S-layer gene (*sap*) has been cloned in two contiguous fragments in *E. coli*. A highly charged sequence was found to be repeated four times in Sap that could be involved in anchoring the capsule to the peptidoglycan.[63]

Summarizing, it is very likely that crystalline surface layers identified on other pathogens of humans, animal and plants (for compilation see refs. 10-12) may be of similar functional relevance as virulence factor.

5.2.2. PARTICIPATION IN MINERAL FORMATION

The S-layer from *Synechococcus* strain GL24 was shown to serve as template for fine grain mineral formation.[64,65] This cyanobacterium was isolated from Fayettville Green Lake, N.Y., a meromitic lake with exceptionally high Ca^{2+} and SO_4^{2-} concentrations. Cells from *Synechococcus* are covered with a hexagonally ordered S-layer lattice. The mineralization process in which the S-layer functions as a regularly structured template starts with adsorption of Ca^{2+}-ions to appropriate sites within the protein lattice. In the presence of high concentrations of SO_4^{2-}-ions, gypsum is formed. When images of unmineralized S-layers were compared with those of mineralized fragments, it could be shown that the diamond-shaped pores in the S-layer became smaller and changed in shape, suggesting that initial binding of ions takes place in the pores. In this sense, the S-layer fulfills a kind of protective function preventing blocking of the porous cell wall. Shedding of S-layer material which was also observed for other organisms involved in biomineralization processes, could be a common process for bacteria to get rid of mineral depositions on their cell surface thereby maintaining basic vital processes such as growth and division as well as nutrient transport. Although the natural pH value of the lake is 7.9 which promotes formation of gypsum, in the course of seasonal warming, alkalization in the close environment of the plasma membrane occurs because of increased photosynthesis. This pushes the mineralization process towards stable calcite crystals in which the sulfate is replaced by carbonate. A similar sulfate to carbonate transformation takes place when Sr^{2+} is the major divalent cation present, forming celestite and strontionite. In experimental systems, to which Sr^{2+} and Ca^{2+} ions were added in equimolar amounts, $CaSr(CO_3)_2$ was formed.[66] This provided evidence that celestite and strontionite, previously thought to be purely evaporitic minerals, can biogenically be formed. It is tempting to speculate that S-layers and bacteria have an important role in the initial development of a range of fine-grain minerals. Because prokaryotes exist since approximately 3.5 billion years, they could have had a major impact on global crust development.

5.2.3. ADHESION OF EXOENZYMES

More recently, S-layers from various Bacillaceae have been suggested or reported to function as an adhesion site for bacterial exoenzymes. The pullulanase of *Thermoanaerobacterium thermosulfurigenes* EM1 was termed pullulanase type II because it hydrolyzes α-1,6 as well α-1,4 linkages in various soluble and branched sugar polymers.[67] Seven major forms of the pullulanase with molecular weights ranging from 93,000 to 490,000 have been identified.[68,69] Sequence analysis revealed a composite structure of the pullulanase consisting of catalytic and noncatalytic domains. The C-terminal half is not necessary for enzymatic function and consists of at least two different segments, separated by a linker region rich in glycine, serine and threonine[69] probably forming a loop.[70] The segment of about 70,000 molecular weight at the N-terminal part of the linker region carries two copies of a fibronectin type III-like domain which has been found in several bacterial depolymerases.[71] The region at the C-terminal half of the linker region contains three repeats of 47, 49 and 51 amino acids which were also found at the N-terminal region of the S-layer proteins from *Acetogenium kivui*,[72] *Thermus thermophilus*[73] and *Thermotoga maritima*.[74] In addition, these S-layer like repeats are also present in the endoxylanase of *Thermoanaerobacter saccharolyticum*,[75] the cellulase of *Bacillus sp.*[76] and the outer layer protein (OlpA) of *C. thermocellum*.[77] Although the nucleotide sequence of the S-layer gene of *T. thermosulfurigenes* EM1 has not yet been elucidated, it was suggested that S-layer homologous domains in the pullulanase could function as an adhesion site for the enzyme.[69] This model would explain why under starch or phosphate limitation, the enzyme is no longer cell associated.[78] Since under these conditions, degradation of the cell envelope including the S-layer occurs, putative attachment sites for the enzyme are lost. For the xylanase of *T. thermosulfurigenes* EM1 a similar mechanism for cell association has been proposed. On the contrary, the α-amylase which is located at the cell periphery, too, does not possess S-layer like domains.

For *B. stearothermophilus* DSM 2358 the putative role of the S-layer as an adhesion site for a high molecular weight enzyme could be confirmed by affinity studies and immunolabeling techniques.[79] The S-layer lattice from this organism completely covers the cell surface and exhibits oblique symmetry (Fig. 5.1). During growth on starch medium, three amylases with molecular weights of 58,000, 98,000 and 184,000 are secreted into the culture fluid but only the high-molecular weight enzyme was found to be associated with whole cells. This enyzme showed affinity to S-layer carrying cell wall fragments and S-layer self-assembly products (Fig. 5.2) but none to isolated peptidoglycan. The surface location of the high-molecular-weight amylase was confirmed by immunogold labeling of whole cells with a polyclonal anti-amylase antiserum (Fig. 5.1). The molar ratio of S-layer subunits (M_r 98,000) to the bound amylase (M_r 184,000) was in the range of 8:1. Such a binding density corresponds to one enzyme molecule per four morphological subunits which guarantees an unhindered transport for nutrients and metabolites through the S-layer lattice. Although it was shown that at least a major

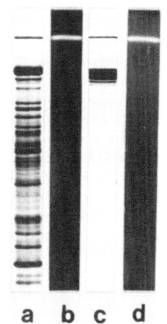

Fig. 5.1. Electron micrographs from (a,b) freeze-etched and (c) negatively-stained preparations from whole cells of Bacillus stearothermophilus DSM 2358. The cell surface (a) before and (b) after extraction of the high-molecular-weight amylase which led to exposure of the oblique S-layer lattice structure. (c) Immunogold-labeling of whole cells with anti-amylase-antiserum for localization of the S-layer bound amylase. Bar in (a,b) 100 nm; in (c) 500 nm.

Fig. 5.2. SDS-PAGE patterns of whole cells (a,b) and S-layer self-assembly products (c,d) from Bacillus stearothermophilus DSM 2358. (a,c) silver staining of protein bands; (b,d) detection of bands with amylolytic activity by applying the iodine-starch reagent. The high-molecular-weight amylase (M, 184,000) that was bound to whole cells (a,b) remained associated with S-layer self-assembly products (c,d).

portion of the amylase is located on the outer face of the S-layer lattice, the question arises as to how the enzyme is associated with the S-layer protein. Several experiments suggest that the amylase is bound to S-layer protein domains either located at the surface of the crystal lattice or exposed inside the pores. The high-molecular weight amylase and the S-layer protein showed no crossreaction with polyclonal antisera indicating the absence of structurally homologous domains.

C. thermocellum produces a highly active thermostable cellulase system in which the various cellulolytic components are associated into a high-molecular-weight complex termed the cellulosome.[80,81] The cellulosome is found in the culture medium and at the cell surface which is covered with a hexagonally ordered S-layer lattice consisting of a 130,000 molecular weight glycoprotein.[82] The catalytic subunits of the cellulosome are organized around a large scaffolding component, termed CipA which was shown to promote binding of the cellulosome to cellulose.[82-86] Sequencing of DNA downstream from *cipA* revealed three open reading frames termed ORF1, ORF2 and ORF3. The C-terminal regions of the polypeptides (ORF1p, ORF2p and ORF3p) encoded by the three open reading frames contain three S-layer-like repeats preceded by a region rich in glycine, proline, threonine and serine. Electron microscopy supported by biochemical investigations revealed that ORF3p[87] and ORF1p[88] are embedded in the same protuberance-forming outer layer of the cell surface. For this reason the genes carrying ORF3 and ORF1 were renamed *olpA* and *olpB* (for outer layer protein), and their products OlpA and OlpB, respectively. The localization of OlpA is in agreement with the hypothesis that the C-terminal triplicated domain of the polypeptide is specific for proteins associated with the S-layer of bacteria.[77] The outer layer of *C. thermocellum* appears to consist of a thick matrix of soft material, shedding particulate fragments into the medium with which OlpA was associated. The role of the OlpA carrying the S-layer-like domain could be to anchor individual cellulases or hemicellulases permanently or transiently to the S-layer of *C. thermocellum*. Recent studies indicated that OlpB might be able to bind non-covalently to the peptidoglycan of *C. thermocellum* by means of its S-layer homologous domains.[88] These observations are consistent with a putative role of S-layer homologous domains in binding to the peptidoglycan layer.[89]

5.3. GENERAL FUNCTIONAL ASPECTS

5.3.1. VARIANT FORMATION

The first indication that variants possessing different S-layer proteins can result from individual clones was reported as early as 1973 by Howard and Tipper.[90] *B. sphaericus* P1 possessing a square S-layer lattice was treated with a virulent bacteriophage for which the S-layer was identified as a binding site. Growth of this organism in the presence of the phage led to the development of 24 resistant S-layer carrying mutants. Interestingly, one group of mutants synthesized comparable amounts of two types of S-layer proteins with different molecular weights.

The diversity of S-layer proteins from *C. fetus subsp. fetus* was originally thought to be a strain-specific feature but changes in S-layer protein synthesis during in vivo or in vitro passage indicated antigenic variation.[43-45,47] A spontaneous S-layer deficient variant 23B was isolated after a laboratory passage of the wild-type strain 23D[91] for which the major S-layer protein showed a molecular weight of 98,000. In order to elucidate molecular mechanisms being involved in antigenic variation, the *sapA* gene[92] of the wild-type strain 23D which encodes the 98,000 S-layer protein was cloned and sequenced. Both, the wild-type strain and the S⁻-mutant contained multiple copies of *sapA* homologs.[44] The formation of the S⁻-phenotype was associated with the failure to produce a transcript for *sapA* or a homologous gene indicating that *sapA* expression is blocked at the transcriptional level. PCR analysis and Southern hydridization indicated that deletion of the promotor region had occurred.[93] Phenotype change in S-layer protein expression from strain 23D to 23D-11 resulting in changes in the molecular weight of the S-layer protein from 98,000 to 127,000 was accompanied by chromosomal rearrangement with site-specific reciprocal recombination.[44] As shown by Southern analysis, the other silent homologs were unaffected by this switch. Strain 23D-11 still showed small amounts of the 98,000 S-layer protein which was similar to observations with other *C. fetus* strains. This may either reflect low level expression from alternative loci or high frequency rearrangements creating mixed populations. Changing of the S-layer protein as the outermost cell envelope component could be advantageous because of protecting the cells against the bactericidal activity of the immune system.

The formation of oxygen dependent variants was observed for *B. stearothermophilus* strains.[94-96] These thermophilic bacteria grow optimally at 60°C and are therefore adapted to low oxygen concentrations only. Thus, cultivation in artificially aerated bioreactors which leads to an increased oxygen tension in the medium and most likely to increased levels of intracellular oxygen radicals may be considered as a kind of stress for the cells.[97] The original S-layer proteins from three different *B. stearothermophilus* strains were synthesized during growth in continuous culture on complex medium only under oxygen limited conditions when glucose was used as the sole carbon source.[94,95] When the rate of aeration was increased up to a level that allowed dissimilation of amino acids as an additional carbon source, the different S-layer proteins from the wild-type strains became replaced by a new common type of S-layer protein with a molecular weight of 97,000 that assembled into an identical oblique (p2) lattice type (Fig. 5.3).[94-96] During the phase of switching from the wild-type strains to the variants, two types of S-layer lattices could be observed on the surface of single cells (Fig. 5.3). The S-layer proteins from the wild-type strains and the variants were structurally different and showed no crossreaction with polyclonal rabbit antisera.[96] Immunogold-labeling of whole cells collected during the phase of switching revealed that the p2-S-layer subunits were randomly inserted into the preexisting hexagonally ordered S-layer lattice from *B. stearothermophilus* PV72. The S-layer deficient variant from strain PV72 "T5" which was isolated by cultivating the wild-type strain for several passages at elevated temperature[98] could express the gene encoding the p2-S-layer protein, too. Although the S-layer proteins from the wild-type strains and the p2-variants were structurally different, the molecular-sieving and the surface properties of the S-layer lattices were not affected by the switch. This indicated that both could represent functionally relevant features of *B. stearothermophilus* strains.[95] Preliminary genetic studies indicated that expression of

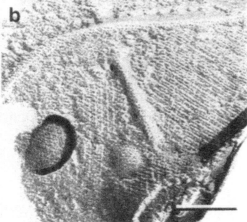

Fig. 5.3. Electron micrographs from freeze-etched preparations from whole cells of Bacillus stearothermophilus PV72 demonstrating oxygen-induced changes in S-layer protein synthesis. Upon increasing oxygen supply in continuous culture, the hexagonally ordered S-layer lattice from the wild-type strain (a) became replaced by an S-layer with oblique lattice symmetry (c). During the phase of switching, both lattice types were detected on the surface of single cells in a monomolecular layer (b). The S-layer protein from the wild-type strain (SbsA) and that from the variant (SbsB) are encoded by different genes. Bars, 100 nm.

the *sbsA* encoding the 130,000 molecular weight S-layer protein from the wild-type strain of *B. stearothermophilus* PV72[99-101] is blocked at the transcriptional level in the p6-deficient variant T5. Furthermore, the failure to detect the *sbsB* gene encoding the 97,000 molecular weight oxygen-induced S-layer protein in the wild-type strain and in the p6-deficient variant T5 indicated genetic rearrangement.[96]

5.3.2. SURFACE PROPERTIES OF S-LAYERS

The most detailed studies on the surface properties of S-layers were performed with mesophilic and thermophilic Bacillaceae.[102-105] Labeling with charged topographical markers and affinity studies revealed that S-layers from several *Bacillus* strains do not possess a net negative surface charge[103-105] as demonstrated for the underlying peptidoglycan-containing layer and for other bacterial cell surface structures such as slimes and capsules.[106,107] Chemical modification of S-layer carrying cell wall fragments from *B. sphaericus* CCM 2120 revealed that per S-layer subunit, 60 carboxyl groups were arranged on surface-located protein domains corresponding to a charge density of 1.6 carboxyl groups per nm^2. In native S-layers, carboxyl groups were neutralized by an equal number of amino groups leading to a charge neutral surface. In peptidoglycan-containing sacculi, only 40 carboxyl groups were arranged at an area comparable to that occupied by one S-layer subunit, confirming that the surface charge density of the S-layer lattice is higher than that of the underlying rigid cell wall layer.[108]

For glycosylated S-layer proteins, the long carbohydrate chains were found to be exposed to the ambient environment.[109] Adsorption studies with whole cells possessing glycosylated S-layer proteins showed that they can bind to hydrophilic, hydrophobic, positively and negatively charged materials to a comparable extent.[102] Cell adhesion of the S-layer carrying *B. stearothermophilus* PV72 was less influenced by the environmental conditions than that of the S-layer deficient T5.[110] Hydrophobic interaction chromatography further revealed a more pronounced hydrophilic surface for the strain lacking the S-layer.[110] For *A. salmonicida* it was demonstrated that the presence of the S-layer makes the cell surface much more hydrophobic.[26] The results available strongly indicate that S-layers are capable of interacting with particles and materials of different physicochemical properties thereby favoring adherence of whole cells to solid surfaces. On the other hand, S-layers from thermophilic Bacillaceae did not adsorb charged macromolecules on their outermost surface or inside the pores[104,108] which would hinder the transport of nutrients and metabolites. Therefore, S-layers can be considered as structures with excellent "antifouling" properties.[108]

5.3.3. MOLECULAR-SIEVE AND PROTECTIVE FUNCTION

S-layers have frequently been suggested to fulfill a protective function for the living cell by excluding hostile lytic enzymes such as muramidases and proteases. However, this assumption could only be confirmed for a few examples. The S-layer from *Sporosarcina urea* protected the murein from lysozyme attack,[111] possibly due to the presence of pores smaller than the 3.5 nm sized enzyme molecules.[112,113] On the other hand, the S-layers from 35 *B. stearothermophilus* strains allowed free passage for the differently sized muramidases lysozyme (M_r 14,600)[114] and mutanolysin (M_r 24,000) indicating that the pores have a size in the range of 4 nm.[115-117] S-layer carrying, muramidase resistant strains of mesophilic *Bacillus* species were found to have a muramidase resistant chemically modified peptidoglycan but did not possess pores significantly smaller than those of strains of thermophilic Bacillaceae.[103] Permeability studies performed on S-layers from *B. stearothermophilus, B. sphaericus, B. alvei, B. coagulans, B. brevis* and *Th. thermohydrosulfuricus* showed that despite different lattice types and different molecular weights of the S-layer subunits, the molecular weight cut off of their S-layers was in the range of 30,000 to 40,000, when structurally well-defined proteins were used as test molecules.[108,115-117]

Since S-layers as isoporous molecular-sieves have the potential to determine the speed of release of the organisms' own exoproteins or even prevent their liberation,[104] it has been suggested that dilute environments could have provided the selection pressure required for developing these crystalline arrays. From the permeability studies it can be assumed that, at least for Gram-positive organisms, S-layers could delineate a compartment outside the cytoplasmic membrane, analogous to the periplasmic space of

Gram-negative eubacteria.[118,119] Based on electron microscopic studies, a similar function was suggested for S-layers of archaeaobacteria lacking a rigid cell wall layer such as halophiles, thermoacidophiles and methanogens. For example, the hexagonally ordered S-layers from *Thermoproteus tenax* and *Th. neutrophilus* have a rather smooth surface with crater-shaped openings,[120,121] whereas the inner face shows prominent protrusions extending towards the cell membrane, probably penetrating it. These protrusions could act as spacer elements, maintaining a 25 nm wide interspace between the cell membrane and the S-layer protein. A similar distance between the hexagonally ordered S-layer lattices and the plasma membrane was observed for *Pyrobaculum organothrophum* strain H10 and *P. islandicum* strain GEO3.[122-124] The long domains projecting from the inner face of the S-layer probably insert into the plasma membrane, which could stabilize the membrane by tethering it to the rigid S-layer lattice. Although there is no real evidence, it was suggested that the space between the S-layer and the cell membrane could contain secreted macromolecules involved in degradation, nutrient transport and folding and export of exoproteins. For the halophiles *Halobacterium halobium* and *Haloferax volcanii* (formerly *Halobacterium volcanii*) a distinct interspace of constant width between the plasma membrane and the S-layer is characteristic, too. The S-layer glycoprotein from *H. halobium* and *H. volcanii* showed a single hydrophobic stretch of 21 amino acids at the C-terminal region which penetrates the plasma membrane and most probably serves as membrane anchor.[125-127] S-layers on Gram-negative eubacteria were found to mask the outer membrane components and receptors, including potential attachment sites for *Bdellovibrio*. *B. bacteriovorus* is an aerobic, obligate predator with biphasic life style. Single, motile attack-phase cells encounter potential prey that are always Gram-negative bacteria by random collosion. The chemical nature of prey-cell identification is still unknown but is not as restrictive as for bacteriophage attachment.[128] Studies on different Gram-negative bacteria such as strains from *Aquaspirillum serpens*, *A. salmonicida*, *C. fetus and Caulobacter crescentus* and different predators showed that not all of the S-layer carrying prey cells were susceptible to the same predator strain.[129,130] A protective function of the S-layer towards *B. bacteriovorus* attack was found for

A. serpens VHA, *A. serpens* MW5 and *A. sinosum*, whereas the isogenic S-layer deficient variant of *A. serpens* VHA, strain VHL, was rapidly attacked. Similar observation that an S-layer protects the cells from *B. bacteriovorus* was done for *C. fetus*. Although many studies on the resistance of S-layer carrying and S-layer deficient strains against grazing protozoa were performed, it is not clear whether S-layers really function as protective coats towards these predators.[130]

5.3.4. SHAPE-DETERMINING AND SHAPE-MAINTAINING FUNCTION

Because of their structural simplicity and from a morphogenetic point of view, it was suggested that S-layer-like structures could have fulfilled a barrier and supporting function as required by self-reproducing systems (progenotes) during the early period of biological evolution.[131,132] Evidence for a shape-determining and shape-maintaining function exists in those archaeaobacteria that synthesize S-layers as the only cell wall component outside the plasma membrane (see chapter 2). In order to fulfill such a function, S-layers must cover the entire cell surface of the bacterium as a closed crystalline container. This requires continuous synthesis of the constituent subunits and their insertion into the crystalline array at defined locations.

The subunits in the hexagonally ordered S-layer lattices from *Thermoproteus*[120] and *Pyrobaculum*[122-124] are most likely covalently linked to each other leading to a strong and rigid structure that may serve as exoskeleton to protect the cells from mechanical and osmotic stress. At the cylindrical part of these rod-shaped cells, the S-layers were found to be of excellent regularity. Lattice faults such as disclinations which are required for complete coverage of the cells were only detected at the cell poles.[120]

The most detailed studies on the shape-determining function of an archaeobacterial S-layer were reported for the lobed cells of *Methanocorpusculum sinense*.[133] The hexagonally ordered S-layer lattice from this organism is composed of glycoprotein subunits with a molecular weight of 92,000 and forms a porous but strongly interconnected network. Although lattice faults are a geometrical necessity on the surface of a closed protein crystal, studies on *M. sinense* strongly indicated that they could play an important role as sites for incorporation of new morphological units, for the formation of the lobed cell structure and in the cell division process.[133] Most likely, the morphogenetic potential observed for the S-layer of *M. sinense* is also valid for other Gram-negative lobed archaeobacteria possessing an S-layer as the exclusive cell wall component.

For the extremely halophilic *H. halobium*, maintenance of the cylindrical shape of the cells was dependent on glycosylation of the S-layer protein with the highly sulfated negatively charged glycan chains and on the presence of high salt concentrations (4 M NaCl) in the environment.[134-136] A comparison of the amino acid sequence of the S-layer glycoprotein from *H. halobium* with that of the moderately halophilic, flat and pleomorphic *H. volcanii*[137] showed that the proteins share 40.5% identity. Regions of high homology arranged in regular patterns which are most probably involved in subunit to subunit interactions were interrupted by stretches of unrelated sequences. The degree of homology dropped strikingly towards the N-terminal end representing the most extracellular domain.[136] Glycosylation of the two S-layer glycoproteins differed markedly: the moderate halophilic showed only 7 instead of 12 glycosylation sites and glucose and galactose were the main sugar components.[138] In *H. halobium*, the uncharged saccharides were replaced by highly sulfated oligosaccharides. This strikingly altered the surface net charge since 120 additional negatively charged groups per S-layer protein are contributed by the sulfated glycans. The drastic structural modification found at the transition from a moderately to an extremely halophilic S-layer glycoprotein seems to have important function to stabilize the protein structure in saturated salt solutions.[136,138]

A shape-maintaining function of the S-layer was also suggested for *Sulfolobus* strains.[2] The hexagonally ordered S-layer lattices from these thermoacidophilic organisms endow the cells with a high degree of structural flexibility, a considerable plasticity and pleomorphism. The S-layer lattice from the strictly anaerobic *D. mobilis* shows square symmetry and represents an open network composed of cross-shaped morphological units which are released when glycerol is added to the cells.[16] The interactions between the subunits within the morphological units are relatively strong in comparison to those between them. Pores passing through the S-layer have a size of even 12 nm which leads to a highly porous network with a porosity in the range of 80%. Together with the low order of the square lattice, this rather indicates a high degree of structural flexibility than rigidity.

5.4. CONCLUSION

In this chapter, specific biological functions of S-layers were described and a survey on general functional aspects as well as on the surface properties was given. From data available, there is clear evidence that no common functional principle exists. S-layers are quite diverse structures—although all of them are organized as two-dimensional protein crystals with geometrically defined arranged functional groups and pores. The expression of different S-layer genes endows the cells with the ability to adapt the cell surface properties to different environmental conditions. For such changes, only a single type of protein must be exchanged which seems to be less complicated and energy-consuming than changing other constituents of supramolecular cell surface structures such as exopolysaccharides for which numerous different enzymes are required. Studies on the pathogenic *C. fetus* and *A. salmonicida*, on the exoenzyme producing Bacillaceae and on the cyanobacterium *Synechococcus* have confirmed that specific biological functions of S-layers can only be elucidated in context with the physiological capabilities and the environmental conditions of the organisms in their natural habitats.

ACKNOWLEDGMENT

This work was supported by the Austrian Science Foundation, project S72/02, and by the Austrian Federal Ministry of Science, Research and the Arts, Republic of Austria.

REFERENCES

1. Baumeister W, Engelhardt H. Three-dimensional structure of bacterial surface layers. In: Harris JR, Horne RW, ed. Electron Microscopy of Proteins Vol 6. Membraneous structures. Academic Press, New York, 1987: 109-54.
2. Baumeister W, Lembcke G. Structural features of archaebacterial cell envelopes. J Bioenerg Biomemb 1992; 24:567-75.

3. Beveridge TJ. Ultrastructure, chemistry and function of the bacterial cell wall. Int Rev Cytol 1981; 72:229-17.

4. Beveridge TJ. Bacterial S-layers. Curr Op Struct Biol 1994; 4:204-12.

5. Beveridge TJ, Graham LL. Surface layers of bacteria. Microbiol Rev 1991; 51:684-705.

6. Beveridge TJ, Koval SF, eds. Advances in Bacterial Paracrystalline Surface Layers. Plenum Press, New York, 1993.

7. Hovmöller S, Sjögren A, Wang DN. The structure of crystalline bacterial surface layers. Prog Biophys Mol Biol 1988; 51:131-63.

8. König H. Archaebacterial cell envelopes. Can J Microbiol 1988; 34:395-406.

9. Koval SF. Paracrystalline protein surface arrays on bacteria. Can J Microbiol 1988; 34:407-14.

10. Messner P, Sleytr UB. Crystalline bacterial cell-surface layers. Adv Microbiol Physiol 1992; 33:213-75.

11. Sleytr UB, Messner P. Crystalline surface layers on bacteria. Annu Rev Microbiol 1983; 37:311-39.

12. Sleytr UB, Messner P, Pum D, Sára M, ed. Crystalline Bacterial Cell Surface Layers. Springer, Berlin, 1988.

13. Sleytr UB, Messner P. Crystalline surface layers in procaryotes. J Bacteriol 1988; 170:2891-97.

14. Sleytr UB, Messner P, Pum D et al. Crystalline bacterial cell surface layers. Mol Microbiol 1993; 10:911-16.

15. Smit J. Protein surface layers of bacteria. In: Inouye M, ed. Bacterial Outer Membranes as Model Systems. Wiley, New York, 1986: 343-76.

16. Wildhaber I, Santarius U, Baumeister W. Three-dimensional structure of the surface protein of *Desulfurococcus mobilis.* J Bacteriol 1987; 169:5563-68.

17. Beveridge TJ, Southam G, Jericho MH et al. High-resolution topography of the S-layer sheath of the archaebacterium *Methanospirillum hungatei* provided by scanning tunneling microscopy. J Bacteriol 1990; 172:6589-95.

18. Beveridge TJ, Sprott GD, Whippey P. Ultrastructure, inferred porosity, and Gram-staining character of *Methanospirillum hungatei* filament termini describe a unique cell permeability for this archaebacterium. J Bacteriol 1991; 173:130-40.

19. Firtel M, Southam G, Harauz G et al. Characterization of the cell wall of the sheathed methanogen *Methanospirillum hungatei* GP1 as an S-layer. J Bacteriol 1993; 175:7550-60.

20. Firtel M, Southam G, Harauz G et al. The organization of the multilayered spacer-plugs of *Methanospirillum hungatei.* J Struct Biol 1994; 112:160-71.

21. Messner P, Sleytr UB. Bacterial surface layer glycoproteins. Glycobiol 1991; 1:545-51.

22. Dooley JSG, Engelhardt H, Baumeister W et al. Three-dimensional structure of the surface layer from the fish pathogen *Aeromonas salmonicida*. J Bacteriol 1989; 171:190-7.

23. Chu S, Cavaignac S, Feutrier J et al. Structure of the tetragonal surface virulence array protein and gene of *Aeromonas salmonicida*. J Biol Chem 1991; 266:15258-65.

24. Gustafson CE, Thomas CJ, Trust TJ. Detection of *Aeromonas salmonicida* from fish by using polymerase chain reaction amplification of the virulence surface array protein gene. Appl Environ Microbiol 1992; 58:3816-25.

25. Munn CB, Ishiguro EE, Kay WW et al. Role of surface components in serum resistance of virulent *Aeromonas salmonicida*. Infect Immun 1982; 36:1069-75.

26. Trust TJ, Kay WW, Ishiguro EE. Cell surface hydrophobicity and macrophage association of *Aeromonas salmonicida*. Curr Microbiol 1983; 9:315-18.

27. Kay WW, Thornton JC, Garduño RA. Structure-function aspects of the *Aeromonas salmonicida* S-layer. In: Beveridge TJ, Koval SF, ed. Advances in Bacterial Paracrystalline Surface Layers. Plenum Press, New York, 1993:151-57.

28. Garduño RA, Phipps BM, Baumeister W et al. Novel structural patterns in divalent cation-depleted surface layers of *Aeromonas salmonicida*. J Struct Biol 1992; 109:184-95.

29. Garduño RA, Phipps BM, Kay WW. Physiological consequences of the S-layer of *Aeromonas salmonicida* in relation to growth, temperature, and outer membrane permeation. Can J Microbiol 1994; 40:622-29.

30. Kay WW, Phipps BM, Ishiguro EE et al. Porphyrin binding by the surface array virulence protein of *Aeromonas salmonicida*. J Bacteriol 1985; 164:1332-36.

31. Phipps BM, Kay WW. Immunoglobulin binding by the regular surface array of *Aeromonas salmonicida*. J Biol Chem 1988; 263:9298-303.

32. Trust TJ, Kostrzynska M, Emödy L et al. High-affinity binding of the basement membrane protein collagen type IV to the crystalline virulence surface protein array of *Aeromonas salmonicida*. Mol Microbiol 1993; 7:593-600.

33. Doig P, Emödy L, Trust TJ. Binding of laminin and fibronectin by the trypsin-resistant major structural domain of the crystalline virulence surface array protein of *Aeromonas salmonicida*. J Biol Chem 1992; 267:43-49.

34. Kokka RP, Vedros NA, Janda JM. Electrophoretic analysis of the surface components of autoagglutinating surface array protein-positive and surface array protein-negative *Aeromonas hydrophila* and *Aeromonas sobria*. J Clin Microbiol 1990; 28:2240-47.

35. Murray RGE, Dooley JSG, Whippey PW et al. Structure of an S-layer on a pathogenic strain of *Aeromonas hydrophila.* J Bacteriol 1988; 170:2625-30.

36. Belland RJ, Trust TJ. Cloning of the gene for the surface protein array of *Aeromonas salmonicida* and evidence linking loss of expression with genetic deletion. J Bacteriol 1987; 169:4086-91.

37. Cover TL, Blaser MJ. The pathobiology of *Campylobacter* infections in humans. Annu Rev Med 1989; 40:269-85.

38. Blaser MJ. Role of the S-layer proteins of *Campylobacter fetus* in serum-resistance and antigenic variation: a model of bacterial pathogenesis. Am J Med Sci 1993; 306:325-29.

39. Blaser M. Biology of *Campylobacter fetus* S-layer proteins. In: Beveridge TJ, Koval SF, ed. Advances in Bacterial Paracrystalline Surface Layers. Plenum Press, New York, 1993: 173-80.

40. Perez-Perez GI, Blaser MJ, Bryner J. Lipopolysaccharide structures of *Campylobacter fetus* related to heat-stable serogroups. Infect Immun 1986; 51:209-12.

41. Yang L, Pei Z, Fujimoto S et al. Reattachment of surface array proteins to *Campylobacter fetus* cells. J Bacteriol 1992; 174:1258-67.

42. Pei Z, Blaser MJ. Pathogenesis of *Campylobacter* infections. Role of surface array proteins in virulence in a mouse model. J Clin Invest 1990; 85:1036-43.

43. Dubreuil JD, Kostrzynska M, Austin JW et al. Antigenic differences among *Campylobacter fetus* S-layer proteins. J Bacteriol 1990; 172:5035-43.

44. Tummuru M, Blaser MJ. Rearrangement of *sapA* homologs with conserved and variable regions in *Campylobacter fetus.* Proc Nat Acad Sci USA 1993; 90:7265-69.

45. Wang E, Garcia M, Blake MS et al. Shift in S-layer protein expression responsible for antigenic variation in *Campylobacter fetus.* J Bacteriol 1993; 175:4979-84.

46. Pei Z, Ellison RT, Lewis RV et al. Purification and characterization of a family of high molecular weight surface-array proteins from *Campylobacter fetus.* J Biol Chem 1988; 263:6416-20

47. Fujimoto S, Takade A, Amako K et al. Correlation between molecular size of the surface array protein and morphology and antigenicity of the *Campylobacter fetus* S-layer. Infect Immun 1991; 59:2017-22.

48. Weiss E, Moulder JW. The rickettsias and chlamydias. Genus I Rickettsia da Rochalima 1916,567. In: Kreig NR, ed. Bergey's Manual of Systematic Bacteriology. vol 1, Williams and Wilkins, Baltimore, 1984:688-98.

49. Carl M, Dasch GA. The importance of the crystalline surface layer protein antigens of rickettsiae in T-cell immunity. J Autoimmun 1989; 2:81-91.

50. Anacker RL, List RH, Mann RE et al. Characterization of monoclonal antibodies protecting mice against *Rickettsia rickettsii*. J Infect Dis 1985; 151:1052-60.

51. Dasch GA, Bourgeois AL. Antigens of the typhus group of rickettsiae: importance of the species-specific surface protein antigens in eliciting immunity. In: Burgdorfer W, Anacker RL, eds. Rickettsiae and Rickettsial Diseases. Academic Press, New York, 1981: 61-70.

52. Carl M, Ching W-M, Dasch GA. Recognition of typhus group rickettsia infected targets by human lymphokine-activated killer cells. Infect Immun 1988; 56:2526-29.

53. Tuszynski GP, Knight LC, Kornecki E et al. Labeling of platelet surface proteins with ^{125}I by the Iodogen method. Anal Biochem 1983; 130:166-70.

54. Kawata T, Takeoka A, Takumi K et al. Demonstration and preliminary characterization of a regular array in the cell wall of *Clostridium difficile*. FEMS Microbiol Lett 1984; 24:323-28.

55. Takeoka A, Takumi K, Kawata T. Purification and characterization of S-layer proteins from *Clostridium difficile* GAI 0714. J Gen Microbiol 1991; 137:261-67.

56. Lortal S, van Heijenoort J, Gruber K et al. S-layer of *Lactobacillus helveticus* ATCC 12046: isolation, chemical characterization and reformation after extraction with lithiumchloride. J Gen Microbiol 1992; 138:611-18.

57. Lortal S. Crystalline surface-layers of the genus *Lactobacillus*. In: Beveridge TJ, Koval SF, ed. Advances in Bacterial Paracrystalline Surface Layers. Plenum Press, New York, 1993: 57-65.

58. Palva A, Kahala M, Vidgren G et al. Characterization of the expression and secretion signals of the *Lactobacillus* S-layer protein gene and their use for heterologous expression. FEMS Microbiol Rev 1993; 12:93

59. Reneiro R, Morelli L, Callegari ML et al. Surface proteins in enteric *Lactobacilli*. Ann Microbiol 1990; 40:83-91.

60. Vidgren G, Palva I, Pakkanen R et al. S-layer protein gene of *Lactobacillus brevis*: cloning by polymerase chain reaction and determination of the nucleotide sequence. J Bacteriol 1992; 174:7419-27.

61. Schneitz C, Nuotio L, Lounatmaa K. Adhesion of *Lactobacillus acidophilus* to avian intestinal epithelial cells mediated by the crystalline bacterial cell surface layer (S-layer). J Appl Bacteriol 1992; 74:290-94.

62. Luckevich M, Beveridge TJ. Characterization of a dynamic S-layer on *Bacillus thuringiensis*. J Bacteriol 1989; 171:6656-67.

63. Etienne-Toumelin I, Sirard JC, Duflot E et al. Characterization of the *Bacillus anthracis* S-layer: cloning and sequencing of the structural gene. J Bacteriol 1995; 177:614-20.

64. Schultze-Lam S, Harauz G, Beveridge TJ. Participation of a cyanobacterial S-layer in fine-grain mineral formation. J Bacteriol 1992; 174:7971-81.

65. Schultze-Lam S, Beveridge TJ. Ultrastructure and chemical characterization of a cyanobacterial S-layer involved in fine-grain mineral formation. In: Beveridge TJ, Koval SF, ed. Advances in Bacterial Paracrystalline Surface Layers. Plenum Press, New York, 1993: 67-75.

66. Schultze-Lam S, Beveridge TJ. Nucleation of celestite and strontianite on a cyanobacterial S-layer. Appl Environ Microbiol 1994; 60:447-53.

67. Antranikian G, Herzberg C, Gottschalk G. Production of thermostable α-amylase, pullulanase, and α-glucosidase in continuous culture by a new *Clostridium* isolate. Appl Environ Microbiol 1987; 53:1668-73.

68. Burchhardt G, Wienecke A, Bahl H. Isolation of the pullulanase gene from *Clostridium thermosulfurogenes* (DSM 3896) and its expression in *Escherichia coli*. Curr Microbiol 1991; 22:91-95.

69. Matuschek M, Burchhardt G, Sahm K et al. Pullulanase of *Thermoanaerobacterium thermosulfurigenes* EM1 (*Clostridium thermosulfurogenes*): molecular analysis of the gene, composite structure of the enzyme, and a common model for its attachment to the cell surface. J Bacteriol 1994; 176:3295-302.

70. Rost B, Schneider R, Sander C. Progress in protein structure prediction? Trends Biochem Sci 1993; 18:120-23.

71. Hansen CK. Fibronectin type III-like sequences and a new domain type in procaryotic depolymerases with insoluble substrates. FEBS Lett 1992; 305:91-96.

72. Peters J, Peters M, Lottspeich F et al. S-layer protein gene from *Acetogenium kivui* : cloning and expression in Escherichia coli and determination of the nucleotide sequence. J Bacteriol 1989; 171: 6307-15.

73. Faraldo MM, de Pedro MA, Berenguer J. Sequence of the S-layer gene of *Thermus thermophilus* HB8 and functionality of its promoter in *Escherichia coli*. J Bacteriol 1992; 174: 7458-62.

74. Engel AM, Cejka Z, Lupas A et al. Isolation and cloning of Ompα, a coiled-coil protein spanning the periplasmic space of the ancestral eubacterium *Thermotoga maritima*. EMBO J 1992; 11:4369-78.

75. Lee Y-E, Lowe SE, Henrissat B et al. Characterization of the active site and thermostability regions of endoxylanase from *Thermoanaerobacterium saccharolyticum* B6A-RI. J Bacteriol 1993; 175:5890-98.

76. Ozaki K, Shikata S, Kawai S et al. Molecular cloning and nucleotide sequence of a gene for alkaline cellulase from *Bacillus* sp. KSM-635. J Gen Microbiol 1990; 136:1327-34.

77. Fujino T, Béguin P, Aubert J-P. Organization of a *Clostridium thermocellum* gene cluster encoding the cellulosome scaffolding protein CipA and a protein possibly involved in attachment of the cellulosome to the cell surface. J Bacteriol 1993; 175:1891-99.

78. Antranikian G, Herzberg C, Mayer F et al. Changes in the cell envelope structure of *Clostridium* sp. strain EM1 during massive production of α-amylase and pullulanase. FEMS Microbiol Lett 1987; 41:193-97.

79. Egelseer E, Schocher I, Sára M et al. The S-layer from *Bacillus stearothermophilus* DSM 2358 functions as an adhesion site for a high-molecular-weight amylase. J Bacteriol 1995; 174:1444-51.

80. Coughlan MP, Non-Nami K, Non-Nami H et al. The cellulolytic enzyme complex of *Clostridium thermocellum* is very large. Biochem Biophys Res Commun 1985; 130:904-09.

81. Lamed R, Setter E, Kenig R et al. The cellulosome: a discrete cell surface organelle of *Clostridium thermocellum* which exhibits separate antigenic, cellulose-binding and various cellulolytic activities. Biotechnol Bioeng Symp 1983; 13:163-81.

82. Lamed R, Bayer EA. The cellulosome concept: exocellular/extracellular enzyme reactor centers for efficient binding and cellulolysis, In: Aubert J-P, Béguin P, Millet J, ed. FEMS Symposium 43. Biochemistry and Genetics of Cellulose Degradation. Academic Press, New York, 1988:101-16.

83. Morag E, Bayer EA, Lamed R. Unorthodox intra-subunit interactions in the cellulosome of *Clostridium thermocellum*: identification of structural transitions induced in the S1 subunit. Appl Biochem Biotechnol 1992; 33:205-17.

84. Salamitou S, Tokatlidis K, Béguin P et al. Involvement of separate domains of the cellulosomal protein S1 of *Clostridium thermocellum* in binding to cellulose and in anchoring of catalytic subunits to the cellulosome. FEBS Lett 1992; 304:89-92.

85. Salamitou S, Raynaud O, Lemaire M et al. Recognition specifity of the duplicated segments present in *Clostridium thermocellum* endoglucanase CelD and in the cellulosome-integrating protein CipA. J Bacteriol 1994; 176:2822-27.

86. Tokatlidis K, Salamitou S, Béguin P et al. Interaction of the duplicated segment carried by *Clostridium thermocellum* cellulases with cellulosome components. FEBS Lett 1991; 291:185-88.

87. Salamitou S, Lémaire M, Fujino T et al. Subcellular Localization of *Clostridium thermocellum* ORF3p, a protein carrying a receptor for the docking sequence borne by the catalytic components of the cellulosome. J Bacteriol 1994; 176:2828-34.

88. Lemaire M, Ohayon H, Gounon P et al. OlpB, a new outer layer protein of *Clostridium thermocellum* and binding of its S-layer like domains to components of the cell envelope. J Bacteriol 1995; 177:2451-59.

89. Lupas A, Engelhardt H, Peters J et al. Domain structure of the *Acetogenium kivui* surface layer revealed by electron crystallography and sequence analysis. J Bacteriol 1994; 176:1224-33.

90. Howard L, Tipper DJ. A polypeptide bacteriophage receptor: modified cell wall protein subunits in bacteriophage-resistant mutants of *Bacillus sphaericus* strain P-1. J Bacteriol 1973; 113:1491-505.

91. McCoy EC, Doyle D, Burda K et al. Superficial antigens of *Campylobacter (Vibrio) fetus*: characterization of antiphagocytic component. Infec Immun 1975; 11:517-25.

92. Blaser MJ, Gottschlich EC. Surface array protein of *Campylobacter fetus*: cloning and gene structure. J Biol Chem 1990; 265:14529-35.

93. Tummuru M, Blaser MJ. Characterization of the *Campylobacter fetus sapA* promotor: evidence that the *sapA* promotor is deleted in spontaneous mutant strains. J Bacteriol 1992; 174:5916-22.

94. Sára M, Sleytr UB. Comparative studies of S-layer proteins from *Bacillus stearothermophilus* strains expressed during growth in continuous culture under oxygen-limited and non-oxygen-limited conditions. J Bacteriol 1994; 176: 7182-89.

95. Sára M, Pum D, Küpcü S et al. Isolation of two physiologically induced variant strains of *Bacillus stearothermophilus* NRS 2004/3a and characterization of their S-layer lattices. J Bacteriol 1994; 176: 848-60.

96. Sára M, Kuen B, Mayer H.F et al. Dynamics in oxygen-induced changes in S-layer protein synthesis and cell wall composition in continuous culture from *Bacillus stearothermophilus* PV72 and the S-layer-deficient variant T5. J Bacteriol 1996; submitted

97. Watson K. Microbial stress proteins. Adv Microbiol Physiol 1990; 31:183-23.

98. Sleytr UB, Messner P, Pum D et al. Struktur und Morphogenese periodischer Proteinmembranen bei Bakterien. Mikroskopie 1982; 39:215-32.

99. Kuen B, Lubitz W, Barton G. Structural and functional analysis of the S-layer protein from *Bacillus stearothermophilus*. In: Beveridge TJ, Koval SF, ed. Advcances in Bacterial Paracrystalline Surface Layers. Plenum Press, New York, 1993:143-50.

100. Kuen B, Lubitz W, Sára M et al. S-layer of *Bacillus stearothermophilus* PV72. In: Beveridge TJ, Koval SF, ed. Advances in Bacterial Paracrystalline Surface Layers. Plenum Press, New York, 1993: 303-06.

101. Kuen B, Sleytr UB, Lubitz W. Sequence analysis of the *sbsA* gene encoding the 130-kDa surface-layer protein of *Bacillus stearothermophilus* strain PV72. Gene 1994; 145:115-20.

102. Sára M, Kalsner, Sleytr UB. Surface properties from the S-layer of *Clostridium thermosaccharolyticum* D120-70 and *Clostridium thermohydrosulfuricum* L111-69. Arch Microbiol 1988; 149:527-33.

103. Sára M, Moser-Thier K, Kainz U et al. Characterization of S-layers from mesophilic bacillaceae and studies on their protective role towards muramidases. Arch Microbiol 1990; 153:209-14.

104. Sára M, Pum D, Sleytr UB. Permeability and charge-dependent adsorption properties of the S-layer lattice from *Bacillus coagulans* E38-66. J Bacteriol 1992; 174:3487-93.

105. Sára M, Sleytr UB. Charge distribution of the S-layer of *Bacillus stearothermophilus* NRS 1536/3c and importance of charged groups for morphogenesis and function. J Bacteriol 1987; 169:2084-89.

106. Costeron JW, Marrie TJ, Cheng KJ. Phenomena of bacterial adhesion. In: Savage DC, Fletcher M, ed. Bacterial Adhesion. Mechanisms and Physiological Significance. Plenum Press, New York, 1985: 3-43.

107. Fletcher M. The physiological activity of bacteria attached to solid surfaces. Adv Microbiol Physiol 1990; 32:53-85

108. Weigert S, Sára M. Surface modification of an ultrafiltration membrane with crystalline structure and studies on interactions with selected protein molecules. J Membrane Sci 1995; 106;147-59.

109. Sára M, Küpcü S, Sleytr UB. Localization of the carbohydrate residue of the S-layer glycoprotein from *Clostridium thermohydrosulfuricum* L111-69. Arch Microbiol 1989; 151:416-20.

110. Gruber K, Sleytr UB. Influence of an S-layer on surface properties of *Bacillus stearothermophilus.* Arch Microbiol 1991; 156:181-85.

111. Beveridge TJ. Surface arrays on the cell wall of *Sporosarcina urea.* J Bacteriol 1779; 139:1039-48.

112. Engelhardt H, Saxton OW, Baumeister W. Three-dimensional structure of the tetragonal surface layer of *Sporosarcina urea.* J Bacteriol 1986; 168:309-17.

113. Stewart M, Beveridge TJ. Structure of the regular surface layer of *Sporosarcina urea.* J Bacteriol 1980; 142:302-09.

114. Messner P, Hollaus F, Sleytr UB. Paracrystalline cell wall surface layers of different *Bacillus stearothermophilus* strains. Int J System Bacteriol 1984; 34:202-10.

115. Sára M, Sleytr UB. Molecular-sieving through S-layers of *Bacillus stearothermophilus* strains. J Bacteriol 1987; 169:4092-98.

116. Sára M, Sleytr UB. Production and characteristics of ultrafiltration membranes with uniform pores from two-dimensional arrays of proteins. J Membrane Sci 1987; 33:27-49.

117. Sára M, Sleytr UB. Membrane biotechnology: Two-dimensional protein crystals for ultrafiltration purposes. In: Rehm H-J, Reed G, ed. Biotechnology Vol 6b. VCH, Weinheim, 1988: 615-36.

118. Graham LL, Beveridge TJ, Nanninga N. Periplasmic space and the concept of periplasm. Trends Biochem Sci 1991; 16:328-29.

119. Breitwieser A, Gruber K, Sleytr UB. Evidence for an S-layer protein pool in the peptidoglycan of *Bacillus stearothermophilus*. J Bacteriol 1992; 174:8008-15.

120. Messner P, Pum D, Sára M, et al. Ultrastructure of the cell envelope of the archaebacteria *Thermoproteus tenax* and *Thermoproteus neutrophilus*. J Bacteriol 1986; 166:1046-54.

121. Wildhaber I, Baumeister W. The cell envelope of *Thermoproteus tenax*: three-dimensional structure of the surface layer and its role in shape maintenance. EMBO J 1987; 6:1475-80.

122. Phipps BM, Huber R, Baumeister W. The cell envelope of the hyperthermophilic archaebacterium *Pyrobaculum organothrophum* consists of two regularly arrayed protein layers: three-dimensional structure of the outer layer. Mol Microbiol 1991; 5:253-65.

123. Phipps BM, Engelhardt H, Huber R et al. Three-dimensional structure of the crystalline protein envelope layer of the hyperthermophilic archaebacterium *Pyrobaculum islandicum*. J Struct Biol 1990; 103:152-59.

124. Phipps BM. Structures of paracrystalline protein layers from the hyperthermophilic archaeobacterium *Pyrobaculum*. In: Beveridge TJ, Koval SF, ed. Advances in Bacterial Paracrystalline Surface Layers. Plenum Press, New York, 1993: 23-33.

125. Lechner J, Sumper M. The primary structure of a procaryotic glycoprotein. Cloning and sequencing of the cell wall glycoprotein gene of halobacteria. J Biol Chem 1987; 262:9724-29.

126. Lechner J and Wieland F. Structure and biosynthesis of procaryotic glycoproteins. Annu Rev Biochem 1989; 58:173-94.

127. Sumper M, Berg E, Mengele R et al. Primary structure and glycosylation of the S-layer protein of *Haloferax volcanii*. J Bacteriol 1990; 172:7111-18.

128. Gray KM, Ruby EG. Intercellular signalling in the *Bdellovibrio* developmental life cycle. In: Dworkin M, ed. Microbial cell-cell interactions. ASM, Washington, 1991; 333-66.

129. Koval SF, Hynes SH. Effect of paracrystalline protein surface layers on predation by *Bdellovibrio bacteriovorus*. J Bacteriol 1991; 173:2244-49.

130. Koval SF. Predation on bacteria possessing S-layers. In: Beveridge TJ, Koval SF, ed. Advances in Bacterial Paracrystalline Surface Layers. Plenum Press, New York, 1993: 85-92.

131. Sleytr UB, Messner P. Self-assembly of crystalline bacterial cell surface layers (S-layers). In: Plattner H, ed. Electron Microscopy of Subcellular Dynamics. CRC, Boca Raton, 1989: 13-31.

132. Sleytr UB, Plohberger R. The dynamic process of assembly of two-dimensional arrays of macromolecules. In: Baumeister W, Vogell W, ed. Electron Microscopy at Molecular Dimensions. Springer, Berlin, 1980: 36-47.

133. Pum D, Messner P, Sleytr UB. Role of the S-layer in morphogenesis and cell division of the archaebacterium *Methanocorpusculum sinense.* J Bacteriol 1991; 173:6865-73.

134. Mescher MF, Strominger JL. Structural (shape-maintaining) role of the cell surface glycoprotein from *Halobacterium salinarium.* Proc Nat Acad Sci 1976; 73:2687-91.

135. Wieland F, Lechner J, Sumper M. The cell wall glycoprotein of *Halobacteria*: structural, functional and biosynthetic aspects. Zbl Bakt Hyg I Abt Orig C3 1982; 161-70.

136. Sumper M. S-layer glycoproteins from moderately and extremely halophilic archaeobacteria. In: Beveridge TJ, Koval SF, ed. Advances in Bacterial Paracrystalline Surface Layers. Plenum Press, New York, 1993:109-117.

137. Cohen S, Shilo M, Kessel M. Nature of the salt dependence of the envelope of a Dead Sea arachaebacterium *Haloferax volcanii.* Arch Microbiol 1991; 156:198-203.

138. Mengele R, Sumper M. Drastic differences in glycosylation of related S-layer glycoproteins from moderate and extreme halophiles. J Biol Chem 1992; 267:8182-89.

CHAPTER 6

BIOTECHNOLOGICAL APPLICATIONS OF S-LAYERS

Margit Sára, Seta Küpcü, Uwe B. Sleytr

6.1. INTRODUCTION

Crystalline bacterial cell surface layers (S-layers) are two-dimensional arrays of proteinaceous subunits with a number of specific features that could be exploited for biotechnological applications (for reviews see refs. 1, 2). (i) S-layers possess pores identical in size and morphology which work in the ultrafiltration range.[3-5] Contrary to isoporous S-layers lattices, ultrafiltration membranes produced from amorphous polymers show a wide pore size distribution.[6,7] (ii) S-layers reveal a high density of functional groups such as amino, carboxylic acid and hydroxyl groups on the outermost surface in well-defined position and orientation which could be used for immobilization of biologically active macromolecules.[1,2] (iii) For immobilization studies, S-layers can be applied in different forms, either as coherent layer bound to a microfiltration membrane or as particulate material. Bound to microporous supports, S-layers can be used as matrix for immunoassays and dipsticks[8] or as enzyme carriers in amperometric[9] or optical[10] biosensors. S-layer microparticles showed excellent stability properties towards shear forces and could be used as "escort particles" in affinity cross-flow filtration.[11,12] (iv) S-layers or S-layer carrying cell wall fragments from Gram-positive eubacteria can be prepared

Crystalline Bacterial Cell Surface Proteins, edited by Uwe B. Sleytr,
Paul Messner, Dietmar Pum, Margit Sára. © 1996 R.G. Landes Company.

in high purity. (v) Processes for growth of S-layer carrying
Bacillaceae in either batch or continuous culture have been developed.[13,14] (vi) The sequence from various S-layer genes is known.[15,16]
Since S-layers are self-assembly systems, the construction of fusion
proteins should allow the combination of a two-dimensional protein crystal with biospecific functional proteins (see chapter 4).

6.2. S-LAYER ULTRAFILTRATION MEMBRANES (SUMS)

By studying the permeability properties of isolated S-layers from
various *Bacillus* strains,[3,4,5] it became evident that these two-dimensional protein crystals function as molecular sieves within the
ultrafiltration range. Unlike S-layer lattices which show a very sharp
exclusion limit, conventional ultrafiltration membranes produced
of amorphous polymers (e.g. cellulose acetate or polysulfone) reveal a wide pore size distribution with pores differing as much as
one order of magnitude. The heteroporous structure of conventional ultrafiltration membranes[6,7] is responsible for the less precise
separation characteristics, the low resolution of protein mixtures,
for the inhomogeneous flux distribution and uneven pore clog-
ging. Because of the presence of pores identical in size and
morphology, S-layers were considered as ideal model systems for
producing the first time isoporous ultrafiltration membranes. S-layer
ultrafiltration membranes (SUMs) are made by depositing either
S-layer self-assembly products or S-layer carrying cell wall fragments on commercial microfiltration membranes with a pore size
of 0.04 to 0.1 μm in a pressure dependent procedure.[4,17,17a] Subsequently, the S-layer protein is crosslinked with glutaraldehyde and
Schiff bases are reduced with sodium borohydride. The nominal
molecular weight cut off of SUMs from S-layers of mesophilic and
thermophilic *Bacillus* strains showed no significant difference and
was in the range of 30,000 to 40,000.[3-5] The porosity of S-layers
lies between 30% to 50% and is therefore significantly higher than
that determined for polymeric ultrafiltration membranes with a
maximum of 10%.

Besides the sharp molecular weight cut off, pores in the native
S-layer lattice were found to be charge neutral, thus preventing
nonspecific adsorption of charged macromolecules which would lead
to pore blocking.[18] After crosslinking the S-layer lattice with

glutaraldehyde, SUMs revealed a net negative charge on the surface and inside the pores.[4,19-21] Quantification of the total number of carboxylic acid groups exposed on the S-layer surface was done for the hexagonally ordered S-layer lattice from *B. stearothermophilus* PV72 and the square S-layer lattice from *B. sphaericus* CCM 2120. For both types of S-layers, about 60 carboxylic acid groups could be determined per constituent protein subunit corresponding to a surface charge density of 1 and 1.6 carboxylic acid groups per nm^2, respectively.[21]

SUMs possessing a high density of free carboxylic acid groups did not adsorb negatively charged macromolecules but strongly interacted with proteins exhibiting a positive net charge under the applied experimental conditions, such as cytochrome c (M_r 12,000; pI 10.8), myoglobin (M_r 17,000; pI 6.8), carbonic anhydrase (M_r 30,000; pI 5.6) or polycationized ferritin (PCF; M_r 440,000; pI 12).[4,18-21]

In ultrafiltration processes, protein adsorption which is considered as the first step in membrane fouling,[22-24] is expressed in terms of flux losses for particle free water after protein filtration.[24] By using ultrafiltration membranes composed of two-dimensional crystalline protein lattices with defined pore size and defined surface charge, it was possible for the first time to determine correlations between the pore size and the net charge of the active filtration layer, the molecular characteristics (dimension, net charge) of adsorbed protein molecules and the flux losses caused by adsorption. For example, filtration of solutions from cytochrome c which has a molecular size of about 3.5 nm and was consequently small enough for passing the 4-5 nm wide pores in the S-layer lattice caused flux losses of up to 80%.[18,20] Despite these considerable flux losses, high resolution electron microscopic studies showed that only a single layer of cytochrome c molecules was adsorbed on the surface and inside the pores of the S-layer lattice.[18,21] On the contrary, flux losses measured after filtration of solutions of PCF were in the range of only 20%. Because of the molecular size of 12 nm, PCF was completely excluded from the pores and formed a closed monolayer on the outer face of the S-layer lattice which did not hinder the passage of the small water molecules through the pores arranged in the S-layer lattice.[18,21]

Unlike SUMs, ultrafiltration membranes from amorphous polymers reveal a wide pore size distribution and blocking of pores does not occur simultaneously. Because of the heteroporous structure of polymeric ultrafiltration membranes, the mechanism of pore blocking is not really understood. It is assumed that irreversibly adsorbed protein layers forming a gel-like structure on the membrane surface are responsible for membrane fouling and flux losses.[22,25-27] By using two-dimensional protein crystals with uniform pores it could be demonstrated that adsorption of almost a single layer of protein molecules inside the pores can cause flux losses in a dimension as they were thought to result from protein gel formation. Although this model was originally proposed by Kim et al[25] for heteroporous polymeric membranes, it could be proved for the first time by using isoporous SUMs.[21]

For changing the rejection and adsorption properties of SUMs, chemical modification reactions were performed under conditions not interfering with the crystalline structure of the S-layer lattice.[19-21] After activation of free carboxylic acid groups with carbodiimide, different low molecular weight nucleophiles leading to either neutral, positively charged, more hydrophilic or more hydrophobic SUMs, were covalently bound to the S-layer lattice. A correlation was observed between the molecular size of attached nucleophiles and the shift of the rejection curve to the lower molecular weight range.[19,20] This was interpreted to result from pore size reduction due to enlargement of modified carboxylic acid groups exposed in well-defined position and orientation on protein domains inside the pores. It was also demonstrated that both, the net charge of the active filtration layer and that of the protein molecules used in filtration experiments determine the solute rejection characteristics of SUMs.[20]

6.3. CONTINUOUS CULTURE OF S-LAYER CARRYING ORGANISMS

For most of the S-layer based technologies developed so far, the growth conditions for the bacteria producing the crystalline arrays are of great importance. Up to now, most knowledge on biomass production of an S-layer carrying organism in continuous culture has accumulated with *B. stearothermophilus* PV72. This thermophilic aerobic organism was originally isolated from a beet sugar extraction plant and was cultivated on complex medium rich on yeast extract, peptone and beef extract.[28] For studying the physiology and parameters that are relevant for biomass and consequently cell wall quality as well as for reducing medium costs, a fully synthetic growth medium was developed in continuous culture[14] by applying the pulse and shift technique.[29]

Originally, S-layer self-assembly products from *B. stearothermophilus* PV72 were used for production of isoporous S-layer ultrafiltration membranes (SUMs).[4] Since the pores in the peptidoglycan meshwork were found to be larger than the 4-5 nm sized pores passing through the S-layer lattice,[3,14] a technology could be developed which allowed the use of S-layer carrying cell wall fragments instead of S-layer self-assembly products for SUM production.[4,21] This alternative procedure omitted the time-consuming extraction and recrystallization process of the S-layer protein. On the other hand, it turned out that by using S-layer carrying cell wall fragments for production of SUMs, quite different aspects of cell wall properties became relevant. First of all, growth conditions for the organism had to be found which led to the accumulation of an S-layer protein pool in the peptidoglycan-containing layer sufficient for generating one complete coverage of the inner face of the rigid cell wall layer.[30] Secondly, it was required that in the course of the cell wall preparation procedure this S-layer protein pool could emerge to form an inner S-layer.[30,31] S-layer protein remaining entrapped within the peptidoglycan-containing layer was found to lead to a significant shift of the rejection curve of SUMs to the lower molecular weight range.[14]

By changing different parameters during growth in continuous culture of *B. stearothermophilus* PV72, the following correlation between growth conditions and cell wall properties could be determined. At low specific growth rates (up to 0.2 h^{-1}), the S-layer protein pool synthesized was sufficient for generating a complete inner S-layer but the autolysine activity relevant for peptidoglycan extension in growing cells was too low for liberating the S-layer protein pool in the course of cell wall preparation. Even during controlled autolysis in the presence of 1 M NaCl the S-layer protein pool was not liberated. At a higher specific growth rate

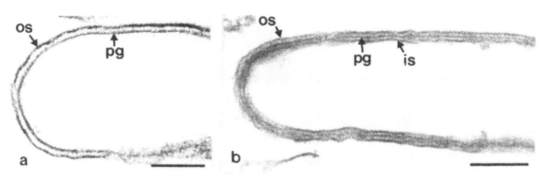

Fig. 6.1. Ultrathin-sections from S-layer carrying cell wall fragments from Bacillus stearothermophilus PV72 *(a) before and (b) after controlled autolysis with 1 M NaCl leading to the liberation of the S-layer protein pool from the peptidoglycan-containing layer of whole cells and to the formation of an inner S-layer. Note that both the outer and the inner S-layer have an identical orientation with regard to the peptidoglycan-containing layer. (os) outer S-layer; (pg) peptidoglycan-containing layer; (is) inner S-layer. Bars, 200 nm.*

$(0.4\ h^{-1})$, the autolysine activity was high but because of limited S-layer protein synthesis, the S-layer protein pool was only sufficient for generating a 30% coverage with an inner S-layer. SUMs produced of such cell wall material did not reveal sufficient long-term stability. Cell wall fragments suitable for high quality SUM production could be obtained at a specific growth rate of $0.3\ h^{-1}$. These cell wall fragments revealed a complete inner S-layer either after the standard cell wall preparation procedure[31] or after mild autolysis[14,30] (Fig. 6.1) induced by defined incubation periods in 1 M NaCl. The peptidoglycan composition, the extent of peptidoglycan crosslinking as well as the amount and composition of secondary cell wall polymers was not influenced by the specific growth rate or the composition of the medium.[14] Standardized procedures in biomass production of S-layer carrying organisms are also very important for preventing changes in S-layer protein synthesis during continuous culture.[32-34]

6.4. S-LAYERS AS MATRIX FOR THE IMMOBILIZATION OF FUNCTIONAL MACROMOLECULES

6.4.1. IMMOBILIZATION PROCEDURES

In eubacterial S-layers the constituent subunits are linked to each other and to the supporting cell envelope components by noncovalent interactions (see chapter 2).[15,16] For obtaining an immobilization matrix stable towards acidic and alkaline pH conditions, against chaotropic agents and organic solvents, S-layer lattices have to be crosslinked with glutaraldehyde before immobilizing functional macromolecules.[35] Another reason for crosslinking the S-layer lattice can be seen in the specific arrangement of functional groups in the two-dimensional protein lattices. The outer surface of native S-layers of *Bacillus* strains reveals an equimolarity of amino and carboxylic acid groups that are involved in direct electrostatic interactions.[18,36,37] For preventing inter- and intramolecular crosslinking reactions within the native S-layer lattice upon activation of carboxylic acid groups or hydroxyl groups which would significantly reduce the immobilization capacity of the crystalline arrays, free amino groups had to be blocked by crosslinking with glutaraldehyde.

For immobilization of enzymes, ligands or antibodies, carboxylic acid groups from the S-layer protein were activated with water-soluble carbodiimide.[35] Activated carboxylic acid groups could then react with free amino groups of the macromolecules leading to stable peptide bonds between the S-layer matrix and the immobilized protein. By applying the carbodiimide activation procedure, it could be demonstrated that the crystalline S-layer matrix favors a controlled immobilization of macromolecules. For example, the large ferritin molecules with a diameter of 12 nm covalently bound to the S-layer clearly reflected the periodicity of the underlying square or hexagonal lattice over a wide range (Fig. 6.2).[1,2,35] For obtaining a covalently crosslinked S-layer lattice with a high density of amino groups, ethylendiamine was linked to carbodiimide-activated carboxylic acid groups.[21] Quantification with butyloxycarbonyl-1-leucine N-hydroxysuccinimide ester (BOC-LEUSE) and amino acid analysis confirmed that the carboxylic acid groups could completely be converted into amino groups, leading to a surface charge density of 1.6 amino groups per nm^2 in case of the square S-layer lattice from *B. sphaericus* CCM 2120 (Fig. 6.2). The advantage of introducing amino groups into the S-layer lattice can be seen in the fact that they are much more reactive than the carboxylic acid groups.

Amino groups introduced into S-layer lattices could be used for immobilization of sulfhydryl group carrying proteins

preactivated with the heterobifunctional crosslinker N-succini-
midyl-3-(2-pyridyldithio)propionate (SPDP) or 3-maleimidobenzoic
acid N-hydroxysuccinimide ester (MBS). On the other hand the
amino groups could subsequently be converted into sulfhydryl
groups by reaction with 2-iminothiolane.[38] For generating cleavable disulfide bonds between the immobilization matrix and the
immobilized protein, sulfhydryl groups introduced into the S-layer
lattice were activated with the homo-bifunctional crosslinker
dipyridyldisulfide (DPDS). For immobilization of macromolecules
to diazoitated amino groups, 2,2 dipyridyl p-phenylendiamine was
first coupled to carbodiimide-activated carboxylic acid groups from
the S-layer protein which was subsequently diazoitated with nitric
acid.[39,40] Although this procedure required rather harsh pH conditions, the structure of the crosslinked S-layer lattice was well
preserved. In case of glycosylated S-layer proteins, macromolecules
could also be linked to the long carbohydrate chains exposed on

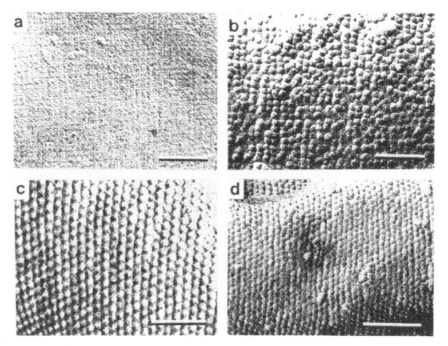

Fig. 6.2. Freeze-dried preparations of cell wall fragments exhibiting the square S-layer
lattice from Bacillus sphaericus CCM 2120 (a) and the hexagonal S-layer lattice from
Thermoanaerobacter thermohydrosulfuricus L111-69 (c) before (a,c) and after (b,d)
covalent binding of ferritin. The immobilized ferritin with a molecular size of 12 nm
reflected the periodicity of the underlying S-layer lattice over a wide range. Bars, 100 nm.

the S-layer surface.[39,41] Vicinal hydroxyl groups were either activated with cyanogen bromide, or if cis-configurations were present, they could be cleaved with periodate. Activation of hydroxyl groups with carbonyldiimidazole was also performed with glycosylated S-layers.

6.4.2. IMMOBILIZATION OF ENZYMES

The different immobilization studies on S-layer lattices showed that regarding the binding density, the retained activity and biospecificity, the optimal activation method strongly depended on the respective enzyme, antibody or ligand.[1,2,35,39] Results from immobilization studies of various enzymes are summarized in Table 6.1. The enzymes were either coupled to the hexagonally ordered S-layer lattices from *Thermoanaerobacter thermohydrosulfuricus* L111-69[42] (Fig. 6.2) or from *B. stearothermophilus* PV72.[28] The covalently bound carbohydrate chains of the S-layer glycoprotein from *Th. thermohydrosulfuricus* L111-69[43,44] could also be exploited for enzyme immobilization.[39,41] Independent of the type of S-layer protein used from different Bacillaceae, the larger enzymes invertase, glucose oxidase, glucuronidase and β-galactosidase formed a monolayer on the outer face of the S-layer lattice (Table 6.1). The enzymatic activity preserved upon immobilization was usually rather high and could be improved to some extent by introducing spacer molecules into the S-layer lattice (Table 6.1). Immobilization of glucose oxidase to the protein moiety of the S-layer lattice from *Th. thermohydrosulfuricus* L111-69 led to significantly better results than covalent binding to the carbohydrate chains (Table 6.1). Because of the high density of either free or succinylated hydroxyl groups, multipoint attachment and entrapping of glucose oxidase molecules between adjacent carbohydrate chains had most likely occurred. This is known to reduce the structural flexibility and to induce irreversible conformational changes, thereby leading to considerable activity loss.[45-47]

The activity of smaller enzymes retained upon immobilization strongly depended on the molecular size of the enzyme, the morphological properties of the S-layer lattice and the applied immobilization procedure.[39] β-glucosidase with a molecular weight of 66,000 was too large to pass through the 4-5 nm wide pores in the S-layer lattice from *Th. thermohydrosulfuricus* L111-69 but was

Table 6.1. Immobilization of various enzymes to the hexagonally ordered S-layer lattice from Thermoanaerobacter thermohydrosulfuricus L111-69

Enzyme	M_r	Immobilization Method	µg Enzyme bound / mg S-layer protein	Molecules / hexameric unit cell	% Retained activity
Invertase	270,000	EDC	1,000	2.6	70
Glucose oxidase	150,000	EDC	600	3.0	35
		Spacer/EDC	600	3.0	60
		BrCN	150	< 1	< 3
		CDI	150	< 1	< 3
		SA/EDC	600	3.0	< 3
β-Galactosidase	116,000	EDC	550	3.4	25
		DPDS	460	2.8	55
Glucuronidase	280,000	EDC	760	2.0	80
Naringinase	96,000	EDC	40	< 1	60
		Spacer/EDC	40	< 1	60
		Diazoitation	280	2.1	80
β-Glucosidase	66,000	EDC	250	2.6	16
		Spacer/EDC	250	2.6	160
Peroxidase	44,000	EDC	350	5.7	< 3
		Spacer/EDC	350	5.7	< 3
		MBS	80	1.3	< 3
		Glutaraldehyde	350	5.7	< 3
		Periodate	300	4.9	< 3

The molecular weight of the glycosylated S-layer subunits is 120,000. Enzymes were immobilized to carbodiimide (EDC)-activated carboxyl groups from the S-layer protein or from 4-amino butyric acid or 6-amino caproic acid introduced into the S-layer lattice as spacer. Hydroxyl groups from the carbohydrate residue of the S-layer glycoprotein were either activated with cyanogen bromide (BrCN), with carbonyldiimidazole (CDI) or with carbodiimide after modification with succinic anhydride (SA). Sulfhydryl groups introduced into the S-layer lattice by modification of amino groups with 2-iminothiolane were activated with the homobifunctional 2,2'-dipyridyldisulfide (DPDS) in the following. Enzymes preactivated with the heterobifunctional m-maleimidobenzoyl-N-hydroxysulfosuccinimide ester (MBS) were bound to amino groups introduced into the S-layer lattice by modification of carbodiimide-activated carboxylic acid groups with ethylenediamine.

small enough for being entrapped in the 6 nm wide funnel-shaped depression in the center of the hexameric unit cells.[39,42,48] When β-glucosidase was linked to carboxylic acid groups from the S-layer protein, only 16% activity could be preserved. A 10-fold increase in activity to even 160% of that from free enzyme was achieved when the enyzme was coupled to spacers such as 4-amino butyric acid or 6-amino caproic acid which increased the distance between the immobilized enyzme and the S-layer lattice. As observed for the larger glucose oxidase linked to the carbohydrate chains, entrapping of the β-glucosidase molecules in cavities of the S-layer protein network which could be prevented by immobilization via spacers obviously reduced the retained enzymatic activity.

The most detailed studies elucidating the significance of interactions between an immobilized enzyme and a defined matrix on activity loss were performed with peroxidase and the hexagonally ordered S-layer lattice from *Th. thermohydrosulfuricus* L111-69.[39] Because of a molecular size of about 4 nm,[49] peroxidase could completely be entrapped inside the 4-5 nm wide pores. Although the enzyme was covalently bound via amino, carboxylic acid or hydroxyl groups which should lead to a different orientation of the active center with respect to the S-layer lattice, in all cases a nearly complete inactivation of the enzyme was observed. The preparation of soluble S-layer protein-peroxidase conjugates in which peroxidase retained at least 40% activity confirmed that not the physicochemical properties of the S-layer protein, but the specific morphological features of the crystalline matrix were responsible for the high activity loss.[39]

Summarizing, S-layers can be considered as ideal immobilization matrix for enzymes which are large enough not to penetrate the pore areas or which can bridge indentations of the corrugated crystal lattice. Because of minor interactions with the immobilization matrix, such large enzymes preserved high activity. On the other hand, enzymes that were small enough to be entrapped in cavities of the S-layer lattice, showed high activity loss or were completely inactivated. For enzymes with a molecular size slightly above the dimension of the pores, this problem could generally be overcome by introducing spacer molecules which prevented intimate contact with the S-layer matrix.

Contrary to S-layers, conventional amorphous and heteroporous polymer materials allow only determination of the influence of the chemical composition of the polymer and of the particular coupling method on activity loss. Even when controlled porous glass was used as an immobilization matrix, the influence of the pore size on the activity loss of immobilized enzymes was not as clear as the impact of the immobilization procedure.[40]

6.4.3. IMMOBILIZATION OF LIGANDS

Protein A (M_r 42,000) is a ligand specific for the Fc-region of most mammalian antibodies leaving the Fab regions free for antigen-specific recognition.[50-52] Since binding of IgG is a reversible process, immobilized Protein A has found many applications in biotechnological downstream-processes. After adsorption of IgG to immobilized Protein A, the Protein A-IgG complex can be cleaved under appropriate conditions such as low pH or high salt concentrations. One major drawback of conventional immobilization matrices for Protein A such as heteroporous beads can be seen in the fact that more than 90% of the ligand is bound in the interior of the sponge-like gel matrix which generally leads to problems because of diffusion controlled reactions.

For immobilization of Protein A to S-layers, the hexagonally ordered lattice from *Th. thermohydrosulfuricus* L111-69 and the square S-layer lattice from *B. sphaericus* CCM 2120 (Fig. 6.2) were selected.[11,12] Protein A was either coupled to carbodiimide-activated carboxylic acid groups from the S-layer protein or was linked to spacer molecules. For immobilization via SPDP, sulfhydryl groups were introduced into Protein A by reaction with 2-iminothiolane. Such modified Protein A still showed about 90% of the original IgG binding capacity and was coupled to amino groups introduced into the S-layer lattice by modification with ethylendiamine. Quantification of Protein A immobilized to carbodiimide-activated carboxylic acid groups showed that up to 700 µg of the ligand was bound per mg S-layer protein from *Th. thermohydrosulfuricus* L111-69 which corresponded to 12 Protein A molecules per hexameric unit cell with a center-to-center spacing of the morphological units of 14.2 nm.[42] Derived from the Stoke'sche radius of this highly elongated protein which was determined to be 5 nm,[50] the saturation capacity of a planar surface lies at 90 ng Protein A/cm². Since the binding capacity of the S-layer lattice for Protein A was found to be in the range of 410 ng/cm² which was significantly above the theoretical value, it can be concluded that for calculating the binding capacity of a planar surface, the real molecule geometry has to be considered.[11] Moreover, most important, it could be demonstrated that the S-layer favored an oriented immobilization of the Protein A molecules with the long axis perpendicularly to the S-layer surface (Fig. 6.3). For determining the IgG binding capacity of immobilized Protein A, polyclonal human IgG was used. The maximum binding capacity was 660 μg IgG per mg S-layer protein which corresponded to a closed monolayer of IgG molecules[11] (Fig. 6.3). Depending on the field of application, Protein A was either linked to the S-layer protein from SUMs which were used for immunoassay and dipstick development,[8] or it was bound to glutaraldehyde crosslinked S-layer carrying cell wall fragments for preparation of affinity microparticles.[12]

6.4.4. AFFINITY MICROPARTICLES

Affinity cross-flow filtration combines a biospecific adsorption process with a membrane separation step.[53,54] After adsorption of target proteins to particulate or macromolecular affinity supports that are retained by an ultrafiltration or a microfiltration membrane, contaminations can be removed by washing with appropriate buffers. Subsequently, conditions such as the pH value or the ionic strength are gradually or immediately changed to impede the affinity interactions. Although affinity cross-flow filtration bears great potential for application in large scale and continuous processes, it is still limited by the availability of suitable "escort" supports.[55] Agarose particles did not resist cross-flow conditions whereas silica particles damaged the delicate surface of the membranes.[55-57]

S-layer carrying cell wall fragments from *Th. thermohydrosulfuricus* L111-69 exhibit a cup-shaped structure with a size of about 1 μm and a complete inner S-layer (Fig. 6.1).[1,2,11,12] After crosslinking the S-layer protein with glutaraldehyde, surface-located free carboxylic acid groups were activated with carbodiimide for immobilization of Protein A.[11] The ligand formed a monolayer on the exposed outer face of the S-layer lattice. Affinity microparticles based on S-layer carrying cell wall fragments revealed excellent stability properties under cross-flow conditions. Even after 24 hour permanent circulation in a cross-flow apparatus equipped with a pinion gear pump, neither the size nor the morphology of the particles had changed and no Protein A leakage or release of S-layer protein could be detected. Since affinity microparticles did not adsorb to cellulose acetate ultrafiltration membranes and could even prevent nonspecific adsorption of IgG by sweeping away reversibly adsorbed protein layers from the membrane surface, they could be used for isolation and purification of IgG from artificial IgG-serum albumin mixtures (Fig. 6.4) or from hybridoma cell culture supernatants.[12] The nominal molecular weight cut off of the cellulose acetate ultrafiltration membrane inserted into the cross-flow

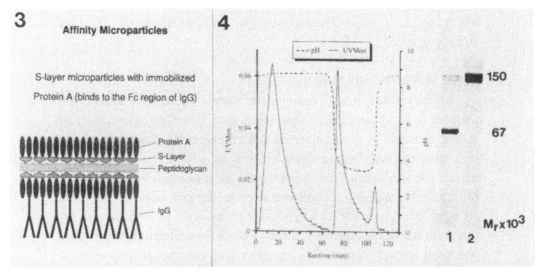

Fig. 6.3. *(Left) Schematic drawing illustrating the mode of immobilization of the elongated Protein A molecules to the outer and inner S-layer of S-layer carrying cell wall fragments. Immobilized Protein A formed a closed monolayer and favoured the correct alignment of subsequently adsorbed IgG molecules.*

Fig. 6.4. *(Right) Affinity cross-flow filtration using Protein A affinity microparticles prepared from S-layer carrying cell wall fragments of* Thermoanaerobacter thermohydrosulfuricus *L111-69. Human polyclonal IgG was isolated from an artificial mixture containing human IgG and serum albumin. The solid line shows the adsorption and elution profile by monitoring the permeate at 280 nm. The first peak obtained during loading of the IgG-serum albumin mixture consisted of non-adsorbed serum albumin. The second peak represented eluted IgG which was obtained after the pH value of the circulating suspension was decreased from 9.0 to 3.5 (dashed line: pH value). As determined by SDS-PAGE, the purity of the eluted IgG was at least 95 %. (lane 1) human IgG-serum albumin mixture; (lane 2) eluted IgG.*

module was in the range of 300,000 which completely rejected the affinity microparticles but allowed free passage for serum albumin in the wash cycle and IgG in the elution step. The main advantage of affinity microparticles in comparison to conventional polymer particles can be seen in their excellent cross-flow stability and immobilization of the ligand on the outermost surface as monomolecular layer which prevented diffusion controlled reactions and nonspecific adsorption of other molecules.

6.4.5. DIAGNOSTIC TEST SYSTEMS

SUMs produced of S-layer carrying cell wall fragments from *B. sphaericus* CCM 2120 showing a square lattice (Fig. 6.2) with a surface charge density of 1.6 carboxylic acid groups per nm² [21] were used as novel matrix for immunoassays and dipsticks.[8] Matrices based on Protein A as an IgG specific ligand were obtained by immobilizing dense monolayers of this ligand to carbodiimide-activated carboxylic acid groups from the S-layer protein of SUMs. For determining the IgG binding capacity, Protein A coated SUMs (PA-SUMs) were incubated with polyclonal human IgG, polyclonal rabbit IgG and monoclonal mouse IgG. After washing the PA-SUMs for removing loosely adsorbed protein, the amount of specifically bound IgG was determined after extraction with SDS by SDS-PAGE. The binding capacity of the PA-SUM for human IgG and rabbit IgG was in the range of 600 ng/cm² and 700 ng/cm² membrane area, respectively, which corresponded to a closed monolayer of IgG molecules on the SUM surface (Fig. 6.5). On the other hand, the binding capacity of the PA-SUM for monoclonal mouse IgG was only 350 ng/cm². This difference can be explained by the lower affinity of mouse IgG to the ligand in comparison to that of human IgG and rabbit IgG.

Alternatively to Protein A, the avidin-biotin or streptavidin-biotin system[58] was applied for SUM-based immunodiagnostic systems. Biotin has high affinity to avidin or streptavidin thereby generating the strongest non-covalent interactions known to date. For obtaining a crystalline protein matrix densely coated with biotin, the square S-layer lattice from *B. sphaericus* CCM 2120 (Fig. 6.2) was crosslinked with glutaraldehyde, and the free carboxylic acid groups were converted into amino groups by

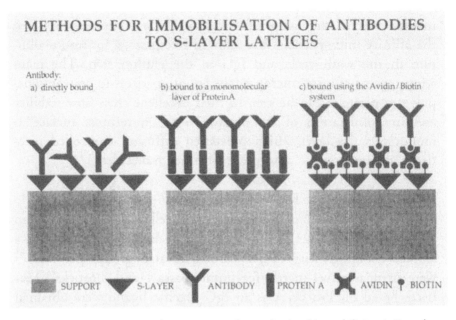

Fig. 6.5. Schematic drawing demonstrating the methods of immobilizing IgG to the S-layer protein in S-layer ultrafiltration membranes (SUMs). Independent on the applied procedure, quantification confirmed that a monolayer of IgG was built up on the S-layer surface.

modification with ethylenediamine.[21] After binding preactivated biotin, such modified S-layers could adsorb up to 800 ng avidin or streptavidin per cm², which corresponded to a closed monomolecular layer (Fig. 6.5).

By immobilizing monolayers of either Protein A or streptavidin to SUMs, a universal biospecific matrix for immunoassays and dipsticks could be generated (Fig. 6.5). Because of the high affinity of rabbit IgG or human IgG to Protein A, the PA-SUM was shown to be particularly suitable for generating dense monolayers of correctly aligned antibodies. On the other hand, mouse IgG with lower affinity to Protein A than human IgG or rabbit IgG was either first biotinylated and was subsequently bound to a streptavidin coated SUM, or it was directly linked to carbodiimide-activated carboxylic acid groups exposed on the surface of the S-layer lattice (Fig. 6.5).

Different detection systems had to be applied when SUMs were used as immobilization matrix for immunoassays or for dipsticks. Quantification of allergens, antigens and antibodies was possible when soluble products were formed by the enzyme-antibody conjugate used in the last step of the detection procedure (Fig. 6.6). For semiquantitative determinations with dipsticks, the detection system involved substrates precipitating on the SUM surface. As an example, quantitative and semiquantitative determination of the main birch pollen allergen Bet$v$Ia[59] is shown in Figure 6.6. In this system, a monoclonal mouse antibody was covalently bound to the S-layer protein. After incubation with different concentrations of Bet$v$Ia, polyclonal rabbit IgG raised against the allergen and subsequently, a goat-anti rabbit alkaline phosphatase conjugate was applied. Quantitative determination of Bet$v$Ia was possible when p-nitrophenyl phosphate was used as detection substrate (Fig. 6.6).

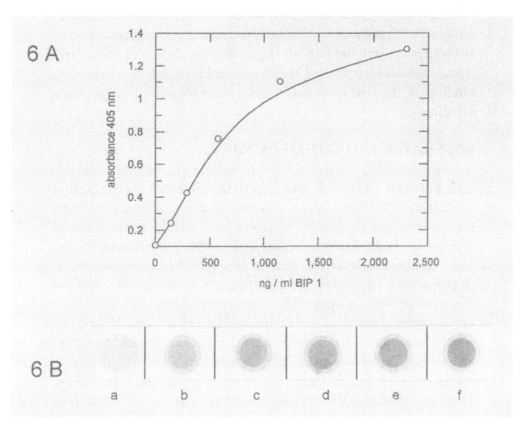

Fig. 6.6. Quantificiation and semiquantiative determination of an allergen after immobilization of a specific monoclonal mouse IgG to the S-layer protein in SUMs. After applying polyclonal rabbit IgG raised against the allergen and a goat-anti-rabbit alkaline phosphatase conjugate, (A) exact quantification was possible when p-nitrophenylphosphate was used as substrate; (B) for semiquantitative determination, BCIP/NBT was used as substrate leading to the formation of a precipitate on the SUM surface. (a) to (e) correspond to the different concentrations of the allergen applied in the immunoassay.

Alternatively, for semiquantitative determination of the bound allergen, 5-bromo-4-chloro-3-indolyl-phosphate/nitro blue tetrazolium (BCIP/NBT) forming a precipitate on the SUM surface was chosen. Most important, SUMs with or without immobilized monoclonal mouse antibody showed no nonspecific adsorption.

The advantage of using SUMs as a matrix for immunoassays and dipstick detection systems can be summarized as follows: (i) the microfiltration membrane is completely covered with a coherent S-layer lattice. Thus, no nonspecific adsorption in the interior of the highly adsorptive polymer matrix of the microfiltration membrane occurs; (ii) in buffer systems, the S-layer surface showed no nonspecific adsorption for foreign proteins; (iii) the S-layer surface is completely covered with a covalently linked dense monolayer of biospecific ligands or antibodies; (iv) since the antigen specific antibodies are immobilized on the outermost surface of the S-layer lattice, diffusion controlled reactions can be excluded; (v) covalent binding of the ligand or antibodies to the S-layer lattice prevents bleeding.

6.5. S-LAYER COATED LIPOSOMES

Artificial lipid vesicles termed liposomes are self-assembly colloid particles in which phospholipid bilayers[60] or tetraether monolayers[61] encapsulate an aqueous medium. Because of their physicochemical properties, liposomes are widely used as model systems for biological membranes and as a delivery system for enhancing the efficiency of various biologically active molecules. Water-soluble molecules can be encapsulated within the aqueous compartment of the liposomes, whereas water-insoluble substances can be solubilized to some extent within the bilayer. Depending on the solubility of the drugs and the fluidity (phase transition characteristics) of the lipid bilayer, such entrapment provides a stable complex, which leads to a longer half life for many drugs. However, the application of liposomes as drug carrier[62] as well as targeting system is limited since the mononuclear phagocytic cells (MPS) of the liver and the spleen are known to remove liposomes from the circulation within a few minutes to hours after intravenous injection. Another problem affecting all particulate drug delivery systems is nonspecific uptake by the phagocytic cells.[63]

On the other hand, the uptake of liposomes by macrophages is responsible for their usefulness as adjuvants for increasing the antigenicity of a variety of molecules.[64]

Regarding the effect of plasma proteins, immunoglobulins and albumin preferably bind to liposomes prepared of phosphatidylcholine. Therefore, substantial progress has been achieved by addition of cholesterol[65] and/or gangliosides, or by the use of solid phase bilayers such as disteroylphosphatidylcholine and sphingomyeline.[66] Macroparticle uptake by macrophages is markedly enhanced when the particle is decorated with a protein ligand that can interact with cell surface receptors on the macrophage. For preventing the normally rapid uptake of the liposomes by the cells of the MPS, so called "stealth" liposomes have been developed.[67] Thereby, monosialogangliosides and more recently, lipid derivatives of polyethylene glycol (PEG) have been identified as stabilizing components. Coating the liposomes with PEG, which led to a more hydrophilic surface,[68] still improved the stealth properties and was found to reduce the opsonization of the liposomes by plasma proteins.[68] If liposomes are applied in gene therapy, the DNA is simply complexed with small unilamellar cationic liposomes. This was found to increase transfection yields by several orders of magnitude.[69]

Recently it was demonstrated that solubilized S-layer protein from *B. coagulans* E38-66[70] can recrystallize onto liposomes composed of dipalmitoylphosphatidylcholine (DP PC), cholesterol and hexadecylamine used in a molar ratio of 10:5:1.[71] The S-layer lattice from this organism showed oblique symmetry with lattice constants of a = 9.4 nm, b = 7.4 nm and γ = 80.[70] The liposomes used for coating with S-layer protein were of the unilamellar type, had a diameter of 100 to 300 nm (Fig. 6.7a) and revealed a positive net charge. As shown by freeze-etching of S-layer coated liposomes (Fig. 6.7b), the subunits had bound with their net negatively charged inner face to the liposomes.

In order to demonstrate the suitability of S-layers bound to liposomes as a matrix for immobilization of functional macromolecules, ferritin with a molecular size of 12 nm was selected which could be visualized by negative-staining and electron microscopic examination procedures.[71] For increasing the stability properties,

Fig. 6.7. Electron micrographs from negatively-stained liposomes composed of dipalmitoylphosphatidylcholine/cholesterol and hexadecylamine (a) before and (b) after recrystallization of the S-layer protein from Bacillus coagulans E38-66 exhibiting oblique (p2) lattice symmetry. (c) Liposomes coated with S-layer protein were used for immobilization of ferritin. Bar in (a) 500 nm; bars in (b) and (c) 50 nm.

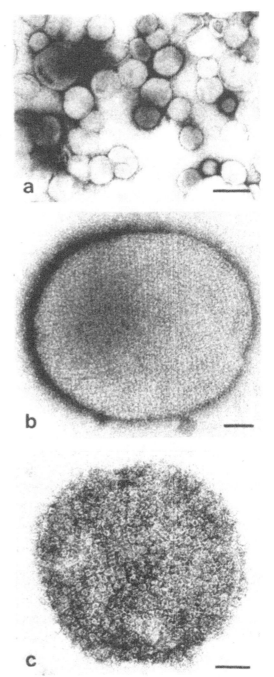

the S-layer protein was crosslinked with glutaraldehyde before carboxylic acid groups from the acidic amino acids were activated with carbodiimide. As confirmed by negative-staining, the large ferritin molecules completely covered the surface of the liposomes (Fig. 6.7c).

Coating the liposomes with a crystalline protein matrix possessing a high potential for chemical modification offers the possibility to adapt the surface properties of the liposomes to the respective requirements in therapeutic use. A further advantage of S-layer coated liposomes can be seen in the existence of a highly ordered immobilization matrix for an oriented anchoring of targeting ligands as well as in the increased stability properties.

6.6. CONCLUSION

S-layers are two-dimensional (glyco)protein crystals composed of identical subunits which possess pores in the ultrafiltration range and reveal functional groups on the outermost surface and in the pore areas in high density, well defined position and orientation. Due to these unique features, they could be exploited for producing the first type of isoporous ultrafiltration membrane. Detailed studies with SUMs led to important knowledge regarding the interactions of a physicochemically well defined porous matrix with differently sized and charged protein molecules. Because of the high density of functional groups on the surface and their accessibility for chemical modification, S-layers are well-defined matrices for a controlled immobilization of functional molecules such as enzymes, antibodies or ligands. After binding as coherent layer to a microporous support, S-layers can be used as crystalline matrix with optimal properties for diagnostic systems (e.g. immunoassays and dipsticks). Prepared as 1 μm large cup-shaped particles, S-layers coated with Protein A were successfully applied as escort particles in affinity cross-flow filtration for isolation of antibodies. Solubilized S-layer subunits have shown the ability to recrystallize onto different templates. This characteristic feature was exploited for coating and stabilizing unilamellar liposomes. The recrystallized S-layers could subsequently be used as regularly structured matrix for immobilizing functional molecules.

Summarizing, S-layers have been optimized in the course of evolution of prokaryotic cells for many specific functions in quite diverse habitats. In comparison to synthetic polymers, this crystalline supramolecular biomaterial reveals significant advantage for a broad spectrum of biotechnological applications.

ACKNOWLEDGMENT

This work was supported by the Austrian Science Foundation, project S72/02, and by the Federal Ministry of Science, Research and the Arts, Republic of Austria.

REFERENCES

1. Sára M, Küpcü S, Weiner C, Weigert S, Sleytr UB. Crystalline protein layers as isoporous molecular sieves and immobilisation matrices. In: Sleytr UB, Messner P, Pum D, Sára M, ed. Immobilised Macromolecules. Application Potentials. Springer, London, 1993:71-86.
2. Sára M, Küpcü S, Weiner C, Weigert S, Sleytr UB. S-layers as immobilisation and affinity matrices. In: Beveridge TJ, Koval SF, ed. Advances in Paracrystalline Bacterial Surface Layers. Plenum Press, New York, 1993:195-204
3. Sára M, Sleytr UB. Molecular-sieving properties through S-layers of *Bacillus stearothermophilus* strains. J Bacteriol 1987; 169:4092-98.
4. Sára M, Sleytr UB. Production and characteristics of ultrafiltration membranes with uniform pores from two-dimensional arrays of proteins. J Membrane Sci 1987; 33:27-49.
5. Sára M, Sleytr UB. Membrane biotechnology: Two-dimensional protein crystals for ultrafiltration purposes. In: Rehm H-J, Reed G, ed. Biotechnology vol 6b. VCH, Weinheim, 1988:615-36.
6. Dietz P, Hansma PK, Inacker O et al. Surface pore structure of micro- amd ultrafiltration membranes imaged with the atomic force microscope. J Membrane Sci 1992; 65:101-11.
7. Marshall AD, Munro PA, Trägardh G. The effect of protein fouling in microfiltration and ultrafiltration on permeate flux, protein retention and selectivity. A literature review. J Membrane Sci 1993; 91:65-108.
8. Breitwieser A, Küpcü S, Weigert S et al. Two-dimensional protein crystals as novel matrix with unique properties for immunoassay and dipstick development. Biotechniques; 1996 submitted.
9. Pum D, Sára M, Sleytr UB. Two-dimensional (glyco)protein crystals as patterning elements and immobilisation matrices for the development of biosensors. In: Sleytr UB, Messner P, Pum D, Sára M, ed. Immobilised Macromolecules. Application Potentials. Springer, London, 1993:142-60.

10. Neubauer A, Pum D, Sleytr UB et al. Fibre-optic glucose sensor using enzyme membranes with two-dimensional crystalline structure. Biosensors Bioelectronics 1995; in press

11. Weiner C, Sára M, Sleytr UB. Novel Protein A affinity matrix prepared from two-dimensional protein crystals. Biotechnol Bioeng 1994; 43:321-30.

12. Weiner C, Sára M, Dasgupta G et al. Affinity cross-flow filtration: purification of IgG with a novel Protein A affinity matrix prepared from two-dimensional protein crystals. Biotechnol Bioeng 1994; 44:55-65.

13. Daugulis AJ, Whitney GK, Wong A et al. Scale-up of S-layer protein secretion by *Bacillus brevis* 47. In: Beveridge TJ, Koval SF, ed. Advances in Paracrystalline Bacterial Surface Layers. Plenum Press, New York, 1993:235-42.

14. Schuster KC, Mayer HF, Kieweg R et al. A synthetic medium for continuous culture of the S-layer carrying *Bacillus stearothermophilus* PV72 and studies on the influence of growth conditions on cell wall properties. Biotechnol Bioeng 1995; 48:66-77.

15. Messner P, Sleytr UB. Crystalline bacterial cell-surface layers. Adv Microbiol Physiol 1992; 33:213-75.

16. Beveridge TJ, Koval SF, ed. Advances in Paracrystalline Surface Layers. Plenum Press, New York, 1993.

17. Sleytr UB, Sára M. Structure with membranes having continuous pores. United States of America Patent 1989; 4,886,604.

17a. Sleytr UB, Sára M. Use of a structure with membranes having continuous pores. United States of America Patent 1989; 4,849,109.

18. Sára M, Pum D, Sleytr UB. Permeability and charge-dependent adsorption properties of the S-layer lattice from *Bacillus coagulans* E38-66. J Bacteriol 1992; 174:3487-93.

19. Küpcü S, Sára M, Sleytr UB. Chemical modification of crystalline ultrafiltration membranes and immobilisation of macromolecules. J Membrane Sci 1991; 61:167-75.

20. Küpcü S, Sára M, Sleytr UB. Influence of covalent attachment of low molecular weight substances on the rejection and adsorption properties of crystalline proteinaceous ultrafiltration membranes. Desal 1993; 90:65-76.

21. Weigert S, Sára M. Surface modification of an ultrafiltration membrane with crystalline structure and studies on interactions with selected protein molecules. J Membrane Sci 1995; 106;147-59.

22. Kim KJ, Fane AG, Fell CJD. Fouling mechanisms of membranes during protein ultrafiltration. J Membrane Sci 1992; 68:79-91.

23. Nilson JL. Protein fouling of ultrafiltration membranes: causes and consequences. J Membrane Sci 1990; 52:121-42.

24. Matthiasson E. The role of macromolecule adsorption in fouling of ultrafiltration membranes. J Membrane Sci 1983; 16:23-36.

25. Kim KJ, Chen V, Fane AG. Some factors determining protein aggregation during ultrafiltration. Biotechnol Bioeng 1993; 42:260-65.

26. Ko MK, Pellegrino JJ, Nassimbene R. Characterization of the adsorption-fouling layer using globular proteins on ultrafiltration membranes. J Membrane Sci 1993; 76:101-20.

27. McDonogh RM, Bauser H, Stroh N et al. Concentration polarisation and adsorption effects in cross-flow ultrafiltration of proteins. Desal 1990; 79:217-31.

28. Sleytr UB, Sára M, Küpcü Z et al. Structural and chemical characterization of S-layers of selected strains of *Bacillus stearothermophilus* and *Desulfotomaculum nigrificans*. Arch Microbiol 1986; 146:19-24.

29. Kuhn H, Friederich U, Fiechter A. Defined minimal medium for a thermophilic *Bacillus* sp. developed by a chemostat pulse and shift technique. Appl Microbiol Biotechnol 1979; 6:341-49.

30. Breitwieser A, Gruber K, Sleytr UB. Evidence for an S-layer protein pool in the peptidoglycan of *Bacillus stearothermophilus*. J Bacteriol 1992; 174:8008-15.

31. Sleytr UB, Glauert AM. Analysis of a regular array of subunits on bacterial surfaces: evidence for a dynamic process of assembly. J Ultrastruct Res 1975; 50:103-16.

32. Sára M, Pum D, Küpcü S et al. Isolation of two physiologically induced variant strains of *Bacillus stearothermophilus* NRS 2004/3a and characterization of their S-layer lattices. J Bacteriol 1994; 176:848-60.

33. Sára M, Sleytr UB. Comparative studies of S-layer proteins from *Bacillus stearothermophilus* strains expressed during growth in continuous culture under oxygen-limited and non-oxygen-limited growth conditions. J Bacteriol 1994; 176:7182-89.

34. Sára M, Kuen B, Mayer HF et al. Dynamics in oxygen-induced changes in S-layer protein synthesis and cell wall composition in continuous culture from *Bacillus stearothermophilus* PV72 and the S-layer-deficient variant T5. J Bacteriol; 1996 submitted.

35. Sára M, Sleytr UB. Use of regularly structured bacterial cell surface layers as matrix for immobilization of macromolecules. Appl Microbiol Biotechnol 1989; 30:184-89.

36. Sára M, Sleytr UB. Charge distribution on the S-layer of *Bacillus stearothermophilus* NRS 1536/3c and importance of charged groups for morphogenesis and function. J Bacteriol 1987; 169:2804-09.

37. Sára M, Sleytr UB. Relevance of charged groups for the integrity of the S-layer lattice from *Bacillus coagulans* E38-66 and for molecular interactions. J Bacteriol 1993; 175:2248-54.

38. Sára M, Sleytr UB. Introduction of sulphhydryl groups into the crystalline bacterial cell surface layer protein from *Bacillus stearothermophilus* PV72 and its application as an immobilization matrix. Appl Microbiol Biotechnol 1992; 38:147-51.

39. Küpcü S, Mader C, Sára M. The crystalline cell surface layer from *Thermoanaerobacter thermohydrosulfuricus* L111-69 as an immobilization matrix: influence of the morphological properties and the pore size of the matrix on the loss of activity of covalently bound enzymes. Biotechnol Appl Biochem 1995; 21:275-86.

40. Manjon A, LLorca FI, Bonhette MJ et al. Properties of β-galactosidase covalently immobilized to glycophase-coated porous glass. Process Biocem 1985; February: 17-22.

41. Sára M, Küpcü S, Sleytr UB. Localization of the carbohydrate residue of the S-layer glycoprotein from *Clostridium thermohydrosulfuricum* L111-69. Arch Microbiol 1989; 151:416-20.

42. Crowther RA, Sleytr UB. An analysis of the fine structure of the surface layers from two strains of *Clostridia*, including correction for distorted images. J Ultrastruct Res 1977; 58:41-49.

43. Christian R, Messner P, Weiner C et al. Structure of a glycan from the surface-layer glycoprotein of *Clostridium thermohydrosulfuricum* strain L111-69. Carbohydr Res 1988; 176:160-63.

44. Bock K, Schuster-Kolbe J, Altman E. Primary structure of the O-glycosidically linked glycan chain of the crystalline surface layer glycoprotein of *Thermoanaerobacter thermohydrosulfuricus* L111-69. J Biol Chem 1994; 269:7137-44.

45. Kennedy JF, Cabral MS. Enzyme Immobilization. In: Rehm H-J, Reed G, ed. Biotechnology vol 7a. Enzyme Technology. VCH, Weinheim, 1987; 347-404.

46. Eddows MJ. Theoretical methods for analysing biosensor performance. In: Cass AEG, ed. Biosensors. A Practical Approach. Oxford University Press 1990; 211-63.

47. Schellenberger A, ed. Enzymkatalyse. Springer, Berlin, 1989.

48. Cejka Z, Hegerl R, Baumeister W. Three-dimensional structure of the surface layer protein of *Clostridium thermohydrosulfuricum*. J Ultrastruct Mol Struct Res 1986; 96:1-11.

49. Tatsuma T, Okawa Y, Watanabe T. Enzyme monolayer- and bilayer modified tin oxide electrodes for the determination of hydrogen peroxide and glucose. Anal Chem 1989; 61:2352-55.

50. Hjelm H, Sjödhal J, Sjöquist J. Immunological activity and structural similar fragments of protein A from *Staphylococcus aureus*. Eur J Biochem 1975; 57:395-403.

51. Faulmann EL. The use of bacterial Fc-binding proteins as probes for antigen-antibody complexes immobilized on nitrocellulose membranes. In: Boyle MDP, ed. Bacterial Immunoglobulin-Binding Proteins: Applications in Immunotechnology, vol 2. Academic Press, London, 1990:250-72.

52. Langone JJ. Protein A of *Staphylococcus aureus* and related immunoglobulin receptors produced by *Streptococci* and *Pneumococci*. Adv Immunol 1982; 32:157-252.

53. Mattiasson B, Ling TGI. Ultrafiltration affinity purification. In: McGregor WC, ed. Membrane Separations in Biotechnology. Marcel Dekker, New York, 1986, 99-114.

54. Mattiasson B, Ramstorp M. Ultrafiltration affinity purification: isolation of Concanavalin A from seeds of *Canavalia ensiformis.* J Chrom 1984; 283:323-30.

55. Herak DC, Merrill EW. Affinity cross flow filtration: some new aspects. Biotechnol Prog 1990; 6:22-40.

56. Luong JHT, Male KB, Nguyen AL. A continuous affinity ultrafiltration process for trypsin purification. Biotechnol Bioeng 1988; 31:516-20.

57. Luong JHT, Male KB, Nguyen AL et al. Mathematical modelling of affinity ultrafiltration processes. Biotechnol Bioeng 1988; 32:451-59

58. Wilchek M, Bayer EA, ed. Avidin-Biotin Technology. Methods Enzymol 1990; 184. Academic Press, Orlando.

59. Breiteneder H, Pettenburger K, Bito A et al. The gene encoding for the major birch pollen allergen, Bet*v* I, is highly homologous to a pea disease resistance gene. EMBO J 1989; 8:1935-41.

60. Lasic DD, Papahadjopoulos D. Liposomes revisited. Science 1995; 267:1275-76.

61. Choquet CG, Patel GB, Beveridge TJ. et al. Formation of unilamellar liposomes from total polar lipid extracts of methanogens. Appl Environ Microbiol 1992; 58:2894-99.

62. Gregoriadis G. ed. Liposomes as Drug Carriers: recent trends and progress. Wiley, New York, 1988.

63. Papahadjopoulos D, Allen T, Garbizon A et al. Sterically stabilized liposomes: improvements in pharmacokinetics, tissue disposition, and anti-tumor therapeutic efficacy. Proc Nat Acad Sci USA 1991; 88:11460-64.

64. Lopez-Berestei G, and Fidler IJ, ed. Liposomes in the therapy of infectious diseases and cancer, Liss, New York, 1989.

65. Kirby C, Clarke J, Gregoriadis G. Effect of the cholesterol content of small unilamellar liposomes on their stability in vivo and in vitro. Biochem J 1980; 186:591-98.

66. Allen TM, Ryan JL, Papahadjopoulos D. Gangliosides reduce leakage of aqueous-space markers from liposomes in the presence of plasma. Biochem Biophys Acta 1985; 818:205-10.

67. Allen TM. Stealth liposomes: avoiding reticuloendothelial uptake. In: Lopez-Berestei G, and Fidler IJ, ed. Liposomes in the Therapy of Infectious Diseases and Cancer, Liss, New York, 1989:405-15.

68. Senior J, Delgado C, Fisher D et al. Influence of surface hydrophilicity of liposomes on their interaction with plasma protein and clearance from the circulation: studies with poly(ethylene glycol)-coated vesicles. Biochim Biophys Acta 1991; 1062:77-82.

69. Zhu N, Liggitt D, Liu Y et al. Systemic gene expression after intravenous DNA delivery into adult mice. Science 1993; 261:209-11.

70. Pum D, Sára M, Sleytr UB. Structure, surface charge, and self-assembly of the S-layer lattice from *Bacillus coagulans* E38-66. J Bacteriol 1989; 171:5269-303.

71. Küpcü S, Sára M, Sleytr UB. Liposomes coated with crystalline bacterial cell surface proteins (S-layer) as immobilization structures for macromolecules. Biochim Biophys Acta 1995; 1235:263-69.

CHAPTER 7

VACCINE DEVELOPMENT BASED ON S-LAYER TECHNOLOGY

Paul Messner, Frank M. Unger, Uwe B. Sleytr

7.1. INTRODUCTION

During the past two decades a considerable body of knowledge has been assembled on ultrastructure, chemistry, and molecular genetics of S-layers[1-7] (see also chapters 2 to 4 of this volume). Bacterial S-layers have been successfully used for the development of isoporous ultrafiltration membranes, as affinity matrices, as supports for Langmuir-Blodgett films and biological membranes, for the production of S-layer-stabilized liposomes, in molecular nanotechnology[8,9] (see also chapters 6 and 8 of this volume) and as an expression system for biotechnologically useful recombinant proteins.[10-12] Other important applications were the use of S-layers for vaccinating fish[13-16] and as carrier/adjuvants for vaccination and immunotherapy.[17-19]

7.2. S-LAYERS AS A FISH VACCINE

Fish represent a major source of food in the world. The increasing world population will continue to demand more fish protein in the future because it is tasty, wholesome and relatively inexpensive to produce. Therefore, fish culture is of increasing importance. Like other animals, fish are susceptible to microbial disease, especially when subjected to intensive culture practices.[14]

Crystalline Bacterial Cell Surface Proteins, edited by Uwe B. Sleytr, Paul Messner, Dietmar Pum, Margit Sára. © 1996 R.G. Landes Company.

In this review, we focus on those fish-pathogenic bacteria which are covered by crystalline S-layers. *Aeromonas salmonicida* and *Aeromonas hydrophila* can cause disease in salmonids in freshwater and marine environments. Typical strains of both species are responsible for furunculosis, a fatal disease of these fishes, while phenotypically 'atypical' strains[20] were isolated from a variety of diseases of many species of freshwater and marine fishes. As was demonstrated by several research groups, the S-layer is essential for virulence of both organisms,[13,21-25] since S-layer-deficient mutants are avirulent.[26,27] Numerous attempts have been undertaken to vaccinate salmon or trout against furunculosis using whole cells, cell sonicates and crude or partially purified extracellular products, but the results reported have been inconsistent.[13,15,28,29] Just recently, a crude acid-extract of an S-layer-positive *A. hydrophila* strain has been used for immunizing channel catfish.[16] When immunization was performed with the S-layer protein-containing extract emulsified in Freund's incomplete adjuvant (FIA), catfish were protected against subsequent experimental challenge with the homologous virulent bacteria. Treatment with FIA only did not protect against the challenge. Upon subsequent immunization with crude acid-extracted S-layer and FIA, the catfish were also protected against challenge with a second, S-layer positive isolate with a different lipopolysaccharide serotype.[16]

7.3. S-LAYERS AS CARRIER/ADJUVANT FOR VACCINATION AND IMMUNOTHERAPY

In conjugate vaccines, the antigens or haptens are bound to a protein by covalent linkages.[30,31] Usually the protein molecules are present as monomers in solution or dispersed as unstructured aggregates. With traditional carriers, reproducible attachment of ligands to carrier proteins is often difficult to achieve. Due to the crystalline nature of S-layer (glyco)proteins, the amino, carboxyl, or hydroxyl groups available for hapten binding occur on each protomer in identical positions and orientations.[32,33] Ligands have been immobilized onto these precisely defined matrices.[8,34] In our experiments, we used the S-layer glycoproteins isolated from *Bacillus stearothermophilus* NRS 2004/3a (henceforth abbreviated as 3a), *Paenibacillus* (formerly *Bacillus*) *alvei* CCM 2051 (2051),

Thermoanaerobacter (formerly *Clostridium*) *thermohydrosulfuricus* L111-69 (L111), and *Thermoanaerobacterium* (formerly *Clostridium*) *thermosaccharolyticum* D120-70 (D120) and the non-glycosylated S-layer protein from *Bacillus stearothermophilus* PV72 (PV72). Two forms of S-layers were prepared for this study, namely, glutaraldehyde-fixed double S-layer sacculi or S-layer self-assembly products (Fig. 7.1a,b). The haptens were coupled either to the protein moiety or to the glycan chains of the respective S-layers[35] (Fig. 7.1a). The S-layer glycoprotein glycans are different from the common eukaryotic glycoproteins (for review see chapter 3 of this volume). In another set of experiments S-layer glycoproteins, which were crosslinked by a cleavable spacer molecule (Fig. 7.1c), were used for coupling of haptens.[36] The latter were immobilized by conventional carbodiimide methodology.[34] All S-layers used in these experiments showed negligible pyrogenicity. Furthermore, no crossreactivity was observed between the S-layer preparations both on the T- or B-cell level (Smith RH, Messner P; unpublished observations).

7.3.1. INDUCTION OF T-CELL IMMUNITY TO TUMOR-ASSOCIATED OLIGOSACCHARIDE ANTIGENS IMMOBILIZED ON S-LAYERS

The identification of certain mucin-associated oligosaccharides as tumor-associated antigens prompted investigations aimed at utilization of these determinants as target haptens for immunotherapy.[37,38] While oligosaccharides per se are too small to effectively engage the immune system, their attachment to carrier proteins results in immunization antigens that are both sufficiently large and capable of involving macrophages and lymphocytes.[30,39] Effective immunotherapeutic responses against tumor-associated antigens are generally assumed to be dependent upon cell-mediated immunity, consisting of CD4+ T-cells that mediate delayed-type hypersensitivity (DTH), CD8+ cytotoxic lymphocytes, and macrophages among other types of cells. Following the investigations of Mosmann and his associates, such cellular responses are mediated primarily by a T-helper cell subset referred to as T_H1.[40] Conversely, such immune responses that depend upon the effector capabilities of antibodies (humoral immune responses)

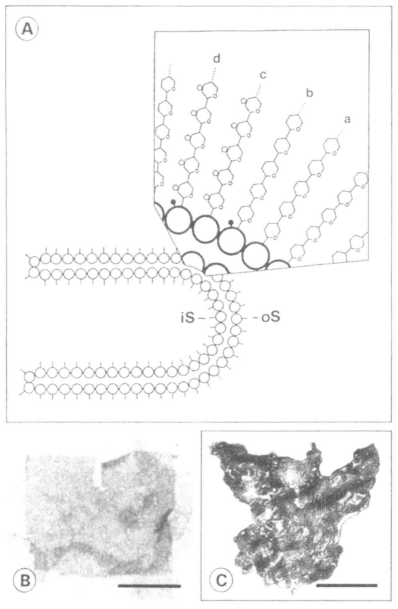

Fig. 7.1. *S-layer preparations used for the immobilization of antigens and haptens. (A) Schematic representation of glutaraldehyde-fixed S-layer conjugates. Functional groups of the protein and carbohydrate moieties, exposed on the inner (iS) and outer (oS) S-layer surface are available for coupling of ligands.*[35] *In the inset, the possibilities for covalent binding of haptens are displayed. (a) Non-activated glutaraldehyde-fixed S-layer glycoprotein. The ligands (●, ○) can be coupled either to the protein moiety (b) or to the glycan chains (c) or to both of them (d). (B) Negatively stained antigen-S-layer conjugate. Sheet-like, double S-layer self-assembly products of* Bacillus sphaericus *CCM 2177 were used as a carrier. Bar = 1 µm. (C) Hapten-S-layer glycoprotein conjugate, crosslinked with a cleavable spacer via the glycan chains. The S-layer glycoprotein was isolated from* Thermoanaerobacter *(formerly* Clostridium*) thermohydrosulfuricus L111-69. Bar = 0.5 µm. (Reproduced from Messner et al. Appl Microbiol Biotechnol 1993; 40:7-11 with permission. © 1993 Springer-Verlag.)*

are mediated primarily by the T_H2 subset. When T-disaccharide αGal1-3βGalNAc (Thomsen-Friedenreich antigen)[37,41] and the Lewis-Y (Le^y) blood group tetrasaccharide αFuc1-2βGal1-4[αFuc1-3]βGlcNAc[42] were used as haptens on S-layer preparations, crosslinked with glutaraldehyde, the following features of the respective immune response were observed:

1. The response was purely a DTH response with practically no antibody production observed. It was found that the quantities of haptenated S-layer preparations required to prime and challenge mice for a strong, hapten-specific DTH response (assayed by determination of footpad swelling) to small carbohydrate haptens are not larger than those needed to generate similar DTH responses to complex protein antigens such as measles or herpes simplex type 1 virus.[43,44] S-layers from different strains, however, showed differences in the DTH responses.

2. Responses were hapten-specific. Following priming with T-haptenated S-layer, only baseline responses were measured when animals were challenged with Le^Y-haptenated S-layer, and vice versa.

3. Strong DTH responses were obtained only in animals that had been pre-treated with cyclophosphamide.

4. The results of an adoptive transfer experiment indicate that T-helper cells are involved in mediating the observed DTH responses.[43]

Groups of mice were pretreated with cyclophosphamide and were then immunized with a preparation of T-disaccharide bound to S-layer from *Bacillus stearothermophilus* NRS 2004/3a (T-3a). Seven days later, the mice were killed, draining lymph nodes and spleens removed, lymphocytes purified and cultured with non-haptenated S-layers from *T. thermohydrosulfuricus* L111-69 (as controls) or with crosslinked T-disaccharide-L111 S-layer conjugate (T-L111). Four days later, all cells were washed with PBS, and the cells stimulated with T-L111 were divided into four groups. One group was sham-treated with PBS, the others, incubated with monoclonal antibodies L3T4, Ly2.2 or Thy1.2. These antibodies were at the time considered directed against markers specific for helper T-cells, suppressor T-cells, and all T-cells (in that order). Incubation of the cell preparations in the presence of the respective antibody and complement was therefore expected to deplete them of T-helper, T-suppressor, or of all T-cells.[43] Following the depletion step, the cells were mixed with either PBS (as control) or with the T-disaccharide-S-layer conjugate with the crosslinked S-layer from *P. alvei* CCM 2051 (T-2051) and injected into naive mice. When T-3a-primed cells stimulated with T-L111 were depleted with the monoclonals L3T4 or Thy1.2 and then mixed with T-2051, these cells did not transfer the potential to generate a DTH response. However, strong DTH responses were observed in mice adoptively transferred with T-3a-primed cells stimulated in vitro with T-L111 and mixed with T-2051. Primed lymphocytes which had been bulk-stimulated with the non-haptenated S-layer L111 alone, then transferred with T-2051, did not enable a DTH response beyond the level of control cells. Moreover, cells treated with antibody Ly2.2 (for depletion of T suppressor cells) also enabled a full DTH response.[43]

Strong DTH responses were also observed in mice primed by oral/nasal application of T-L111 conjugates. The responses are comparable to those achieved following intramuscular immunization.[43]

Presently, a program of chemical synthesis has been initiated to prepare a complete set of mucin-derived, tumor-associated oligosaccharides. These and their chemically modified analogs will be attached to S-layers and examined for their ability to induce a predominantly T_H1-controlled immune response against tumor-associated mucins in humans.

7.3.2. ELICITATION OF IMMUNE RESPONSES
AGAINST S-LAYER CONJUGATES
WITH *STREPTOCOCCUS PNEUMONIAE* CAPSULAR MATERIAL

Successful vaccination against *S. pneumoniae* requires elicitation of protective antibody titers specific for the capsular polysaccharides of these pathogenic organisms. To date, 83 serotypes have been reported.[45] While many of the 83 capsular polysaccharides elicit protective immunity in healthy adult vaccinees, several of them elicit poor responses in infants or immune-compromised patients. Thus, a typical anti-pneumococcal vaccine contains capsular polysaccharide material from 23 serotypes of *S. pneumoniae* (for a review, see ref. 46). To overcome these difficulties, *S. pneumoniae* type 8 was chosen as a model target.[47] Type 8 polysaccharide material, which is used in the commercially available pneumonia vaccine, did not elicit bactericidal antibody responses in mice in our hands. Therefore, oligosaccharides (one to eight repeat units) derived from type 8 capsular polysaccharide by controlled, acid-catalyzed hydrolysis were coupled to non-crosslinked S-layer from *P. alvei* CCM 2051 and injected into mice. Conjugates comprising relatively small oligosaccharide haptens elicited good antibody responses as determined by enzyme immunoassay. Serum from the mice thus immunized had immunoprotective properties as revealed in a serum killing assay.[47] On blood agar plates, the mouse sera reduced *S. pneumoniae* serotype 8 colony forming units by 99%. No reduction was observed with sera from mice that had been immunized with type 8 capsular polysaccharide only. However, injection of conjugates into mice that had previously been immunized with capsular polysaccharide produced sera capable of reducing colony forming units by 65%. In rabbits, results were similar, except for the finding that unconjugated type-8-oligosaccharide alone elicits a level of IgG response somewhat below that of the oligosaccharide-2051 conjugates.[47]

The type 8-oligosaccharide-2051-conjugates also elicited DTH in mice, as shown by the following experiment. Mice were primed intramuscularly with type 8-oligosaccharide-2051 conjugate, the oligosaccharide alone, or with heat-killed *S. pneumoniae* serotype 8. Unhaptenated S-layer or phosphate-buffered saline were used as vaccination controls. One week later, all mice were footpad-challenged with heat-killed *S. pneumoniae* type 8. The results indicate that the conjugates are as effective in priming for a DTH response as the heat-killed bacteria, whereas the oligosaccharide alone is a considerably less effective primer.[47]

Presently, specialized composite vaccines are being designed for high-risk target groups such as infants or immune-compromised patients. Such vaccines will contain oligosaccharide-S-layer conjugates with non-crosslinked S-layers admixed with purified capsular polysaccharides. S-layer oligosaccharide conjugates will be developed for those target organisms whose capsular polysaccharides afford poor vaccination responses in the respective target population.

7.3.3. S-LAYER CONJUGATES FOR SUPPRESSION OF AN IgE RESPONSE TO BETv1, THE MAJOR ALLERGEN OF BIRCH POLLEN

Results of immunization experiments with S-layer conjugates (non-crosslinked or crosslinked S-layers) have indicated that immune responses in animals can be modulated toward a T_H1- or a T_H2-directed response mode through the choice and construction of the respective S-layer conjugates.[19] In the case of small tumor-associated oligosaccharides, crosslinked S-layers effectively induced DTH (presumably a T_H1-directed response) whereas with the *S. pneumoniae* type 8 polysaccharide, non-crosslinked S-layers were more effective in eliciting an immunoprotective antibody response.

Betv1 is the major pollen allergen of the birch and is responsible for atopic (IgE-mediated) allergies in an increasing percentage of the population.[48] After sequencing and cloning, recombinant Betv1[49] was available for the coupling experiments. Presently, different types of Betv1-S-layer conjugates are being studied to determine whether they are capable of converting a (T_H2-directed) IgE antibody response into a T_H1-mediated response against Betv1, indicating an ability to suppress the manifestations of allergy in patients susceptible to pollen allergies.

These studies have only recently been initiated with different S-layer-Betv1 conjugates using both non-crosslinked and crosslinked S-layers of *Bacillus sphaericus* CCM 2177 (Fig. 7.1b) and *Thermoanerobacter thermohydrosulfuricus* L111-69 (Jahn-Schmid B, Graninger M, Glozik M et al. Immunotechnology; submitted). Preliminary results indicate that the allergen-S-layer conjugates are capable of eliciting antibodies in mice and of stimulating human T-cell clones specific for certain Betv1 epitopes.[50]

7.4. CONCLUSIONS

Bacterial surface layer proteins have been utilized in different ways in vaccine development. In the case of fish vaccines to fight

Aeromonas infections in salmonids, the crystalline cell surface protein (referred to as A-layer) itself was considered a good vaccine candidate because of its surface location and its immunological properties. However, the results of the vaccinations are not yet completely satisfying. While in several vaccination models with different S-layer preparations rather distinct levels of protection were observed upon challenge,[15,28,29] obviously better results have been obtained in the vaccination of catfish with acid-extracted crude S-layers.[16] To develop an effective, inexpensive fish vaccine, further research is necessary.

Another approach for using S-layers for vaccine development has been suggested by our group several years ago.[17,18] Due to their crystalline nature, the protein arrays contain functional groups in precisely defined orientations for coupling of haptens. Conventional applications of S-layer conjugates do not cause observable trauma or side effects. Depending on the nature of the S-layer, the antigenic conjugates elicit either predominantly cellular or predominantly humoral immune responses.[43,47] Even oral/nasal administration of S-layer-hapten conjugates does induce significant vaccination responses. One project is directed to immunotherapy of cancer, where conjugates of S-layer with small, tumor-associated oligosaccharides have been found to elicit hapten-specific DTH responses.[43] In another application, oligosaccharides derived from capsules of *Streptococcus pneumoniae* type 8 have been linked to S-layer proteins and have been found to elicit protective antibody responses in animals.[47] Most recently, allergen-S-layer conjugates have been prepared with the intention to suppress the T_H2-directed, IgE-mediated allergic responses to Bet$v$1, the major allergen of birch pollen.[50] In all three applications, the S-layer vaccine technology appears to offer the versatility needed to direct vaccination responses toward predominant control by T_H1 or T_H2 lymphocytes to meet the different therapeutic or prophylactic requirements in each case.

ACKNOWLEDGMENTS

We appreciate the instrumental contributions of Dr. Beatrice Jahn-Schmid to the project and thank Profs. D. Kraft, O. Scheiner and Dr. C. Ebner for valuable discussions and for the hospitality of their facilities at the General Hospital in Vienna. This work was supported by a grant from the Austrian Science Foundation, project S7206-MOB and the Austrian Federal Ministry of Science, Research and the Arts.

REFERENCES

1. Messner P, Sleytr UB. Crystalline bacterial cell-surface layers. In: Rose AH, ed. Advances in Microbial Physiology, Vol. 33. London: Academic Press, 1992:213-75.
2. Sleytr UB, Messner P, Pum D et al. Crystalline bacterial cell surface layers. Mol Microbiol 1993; 10:911-16.
3. Peyret JL, Bayan N, Joliff G et al. Characterization of the *cspB* gene encoding PS2, an ordered surface-layer protein in *Corynebacterium glutamicum*. Mol Microbiol 1993; 9:97-109.
4. Beveridge TJ. Bacterial S-layers. Curr Opin Struct Biol 1994; 4:204-12.
5. Blaser MJ, Wang E, Tummuru MKR et al. High-frequency S-layer protein variation in *Campylobacter fetus* revealed by *sapA* mutagenesis. Mol Microbiol 1994; 14:453-62.
6. Kuen B, Sleytr UB, Lubitz W. Sequence analysis of the *sbsA* gene encoding the 130-kDa surface layer protein of *Bacillus stearothermophilus* strain PV72. Gene 1994; 154:115-20.
7. Noonan B, Trust TJ. Molecular characterization of an *Aeromonas salmonicida* mutant with altered surface morphology and increased systemic virulence. Mol Microbiol 1995; 15:65-75.
8. Sleytr UB, Sára M, Messner P et al. Application potential of 2D protein crystals (S-layers). Ann New York Acad Sci 1994; 745:261-69.
9. Küpcü S, Sára M, Sleytr UB. Liposomes coated with crystalline bacterial cell surface protein (S-layer) as immobilization structures for macromolecules. Biochim Biophys Acta 1995; 1235:263-69.
10. Udaka S, Tsukagoshi N, Yamagata H. *Bacillus brevis*, a host bacterium for efficient extracellular production of useful proteins. Biotechnol Gen Eng Rev 1989; 7:113-46.
11. Udaka S, Yamagata H. High-level secretion of heterologous proteins by *Bacillus brevis*. Methods Enzymol 1993; 217:23-33.
12. Ichikawa Y, Yamagata H, Tochikubo K et al. Very efficient extracellular production of cholera toxin B subunit using *Bacillus brevis*. FEMS Microbiol Lett 1993; 111:219-24.
13. Udey LR, Fryer JL. Immunization of fish with bacterins of *Aeromonas salmonicida*. Marine Fisheries Rev 1978; 40:12-17.
14. Trust TJ. Pathogenesis of infectious diseases of fish. Annu Rev Microbiol 1986; 40:479-502.

15. Evenberg D, De Graaff P, Lugtenberg B et al. Vaccine-induced protective immunity against *Aeromonas salmonicida* tested in experimental carp erythrodermatitis. J Fish Disease 1988; 11:337-50.

16. Ford LA, Thune RL. Immunization of channel catfish with a crude, acid-extracted preparation of motile aeromonad S-layer protein. Biomed Lett 1992; 47:355-62.

17. Sleytr UB, Mundt W, Messner P et al. Immunogenic compositions containing ordered carriers. 1991; US Patent No. 5,043,158.

18. Sleytr UB, Mundt W, Messner P. Pharmazeutische Struktur mit an Proteinträger gebundenen Wirkstoffen. 1993; Eur. Patent No. 0306473.

19. Malcolm AJ, Messner P, Sleytr UB et al. Crystalline bacterial cell surface layers (S-layers) as combined carrier/adjuvants for conjugate vaccines. In: Sleytr UB, Messner P, Pum D et al, eds. Immobilised Macromolecules: Application Potentials. London: Springer, 1993:195-207.

20. Belland RL, Trust TJ. DNA:DNA reassociation analysis of *Aeromonas salmonicida*. J Gen Microbiol 1988; 134:307-15.

21. Kay WW, Buckley JT, Ishiguro EE et al. Purification and disposition of a surface protein associated with virulence of *Aeromonas salmonicida*. J Bacteriol 1981; 147:1077-84.

22. Evenberg D, Lugtenberg B. Cell surface of the fish pathogenic bacterium *Aeromonas salmonicida*. II. Purification and characterization of a major cell envelope protein related to autoagglutination, adhesion and virulence. Biochim Biophys Acta 1982; 684:249-54.

23. Ford LA, Thune RL. S-layer positive motile aeromonads isolated from channel catfish. J Wildlife Dis 1991; 27:557-61.

24. Sakata T, Shimojo T. Surface structure and pathogenicity of *Aeromonas hydrophila* strains isolated from diseased and healthy fish. Memoirs of the Faculty of Fisheries, Kagoshima University 1991; 40:47-58.

25. Fernández AIG, Pérez MJ, Rodríguez LA et al. Surface phenotypic characteristics and virulence of Spanish isolates of *Aeromonas salmonicida* after passage through fish. Appl Environ Microbiol 1995; 61:2010-12.

26. Ishiguro EE, Kay WW, Ainsworth T et al. Loss of virulence during culture of *Aeromonas salmonicida* at high temperature. J Bacteriol 1981; 148:333-40.

27. Dooley JSG, Lallier R, Trust TJ. Antigenic structure of *Aeromonas hydrophila*. Vet Immunol Immunopathol 1986; 12:339-44.

28. Michel C. Progress towards furunculosis vaccination. In: Roberts RJ, ed. Microbial Diseases of Fish. New York: Academic Press, 1982:151-71.

29. Cipriano RC. Furunculosis: pathogenicity, mechanism of bacterial virulence and the immunological response of fish to *Aeromonas salmonicida*. In: Crosa JH, ed. Bacterial and Viral Diseases of Fish. Seattle: University of Washington, 1983:41-60.

30. Schneerson R, Robbins JB, Szu SC et al. Vaccines composed of polysaccharide-protein conjugates: current status, unanswered questions, and prospects for the future. In: Bell R, Torrigiani G, eds. Towards Better Carbohydrate Vaccines. Chichester: Wiley, 1987:307-32.

31. Dick WE Jr, Beurret M. Glycoconjugates of bacterial carbohydrate antigens. In: Cruse JM, Lewis RE Jr, eds. Conjugate Vaccines. Basel: Karger, 1989:48-114.

32. Sára M, Küpcü S, Weiner C et al. S-layers as immobilization and affinity matrices. In: Beveridge TJ, Koval SF, eds. Advances in Bacterial Paracrystalline Surface Layers. New York: Plenum, 1993:195-204.

33. Sára M, Küpcü S, Weiner C et al. Crystalline protein layers as isoporous molecular sieves and immobilisation and affinity matrices. In: Sleytr UB, Messner P, Pum D et al, eds. Immobilised Macromolecules: Application Potentials. London: Springer, 1993:71-86.

34. Sára M, Sleytr UB. Use of regularly structured bacterial cell envelope layers as matrix for the immobilisation of macromolecules. Appl Microbiol Biotechnol 1989; 30:184-89.

35. Messner P, Mazid MA, Unger FM et al. Artificial antigens. Synthetic carbohydrate haptens immobilized on crystalline bacterial surface layer glycoproteins. Carbohydr Res 1992; 233:175-84.

36. Messner P, Wellan M, Kubelka W et al. Reversible cross-linking of crystalline bacterial surface layer glycoproteins through their glycan chains. Appl Microbiol Biotechnol 1993; 40:7-11.

37. MacLean GD, Bowen-Yacyshyn MB, Samuel J et al. Active immunization of human ovarian cancer patients against a common carcinoma (Thomsen-Friedenreich) determinant using a synthetic carbohydrate antigen. J Immunother 1992; 11:292-305.

38. MacLean GD, Reddish M, Koganty R et al. Immunization of breast cancer patients using a synthetic sialyl-Tn glycoconjugate plus Detox adjuvant. Cancer Immunol Immunother 1993; 36:215-22.

39. Jennings HJ. Capsular polysaccharides as human vaccines. Adv Carbohydr Chem Biochem 1983; 41:155-208.

40. Mosmann TR, Coffman RL. Heterogeneity of cytokine secretion patterns and functions of helper T cells. Adv Immunol 1989; 46:111-47.

41. Ratcliffe RM, Baker DA, Lemieux RU. Synthesis of the T-antigenic determinant in a form useful for the preparation of an effective artificial antigen and the corresponding immunoadsorbent. Carbohydr Res 1981; 93:35-41.

42. Hindsgaul O, Norberg T, LePendu J et al. Synthesis of type 2 human blood group antigenic determinants. The H, X and Y haptens and variations of the H type 2 determinants as probes for the combining site of the lectin 1 of *Ulex europaeus*. Carbohydr Res 1982; 109:109-42.

43. Smith RH, Messner P, Lamontagne LR et al. Induction of T-cell immunity to oligosaccharide antigens immobilized on crystalline bacterial surface layers (S-layers). Vaccine 1993; 11:919-24.

44. Smith RH, Ziola, B. Cyclophosphamide and dimethyl dioctadecyl ammonium bromide immunopotentiate the delayed-type hypersensitivity response to inactivated enveloped viruses. Immunology 1986; 58:245-50.

45. Eskola J, Käyhty H. New vaccines for prevention of pneumococcal infections. Ann Med 1995; 27:53-56.

46. Shapiro ED. Pneumococcal Vaccine. In: Cryz SJ Jr, ed, Vaccines and Immunotherapy. New York: Pergamon Press, 1991:127-39.

47. Malcolm AJ, Best MW, Szarka RJ et al. Surface layers from *Bacillus alvei* as a carrier for a *Streptococcus pneumoniae* conjugate vaccine. In: Beveridge TJ, Koval SF, eds. Advances in Bacterial Paracrystalline Surface Layers. New York: Plenum, 1993:219-33.

48. Breiteneder H, Pettenburger K, Bito A et al. The gene coding for the major birch pollen allergen Bet*v*I, is highly homologous to a pea disease resistance response gene. EMBO J 1989; 8:1935-38.

49. Ferreira FD, Hoffmann-Sommergruber K, Breiteneder H et al. Purification and characterization of recombinant Bet*v*I, the major birch pollen allergen. Immunological equivalence to natural *Betv* I. J Biol Chem 1993; 268:19574-80.

50. Jahn-Schmid B, Messner P, Unger FM et al. Toward selective elicitation of T_H1-controlled vaccination responses: vaccine applications of bacterial surface layer proteins. J Biotechnol, in press.

CHAPTER 8

MOLECULAR NANOTECHNOLOGY AND BIOMIMETICS WITH S-LAYERS

Dietmar Pum, Uwe B. Sleytr

8.1. INTRODUCTION

The functionalization of surfaces and interfaces with mono-molecular layers of two-dimensional arrays of biopolymers has already attracted much attention in the field of supramolecular engineering since it offers a feasible way for the fabrication of new bioelectronical devices and advanced materials (for reviews on various model systems see refs. 1-6). In particular, one of the most important objectives is the development of general strategies for the crystallization of proteins on suitable substrates. Many functional proteins require a fluid environment of a cell membrane in order to maintain their activity. Frequently the lifetime of such biosystems is dramatically decreased when artificial surfaces are used instead of biocompatible ones. Thus, new concepts have to be developed in which proteins maintain their function by stabilizing them in a two-dimensional crystalline arrangement.[7]

A new approach in nanostructure technology particularly in the functionalization of surfaces has been developed in our institute on the basis of crystalline bacterial cell surface layers

(S-layers).[8-13] S-layers composed of monomolecular arrays of identical protein or glycoprotein species (see also chapter 2 to 4) have been shown to be ideal patterning structures for supramolecular engineering due to their high molecular order, defined mass distribution and isoporosity, high binding capacity and the ability to recrystallize with perfect uniformity over large areas on solid supports or interfaces including Langmuir-Blodgett (LB) films. It is their simple construction principle, involving a single constituent subunit, which allows a deeper understanding of the molecular logic involved in the generation of these two-dimensional arrays. Figure 8.1 summarizes the resolution capabilities of several optical instruments and the milestones in microfabrication techniques in the course of history. From this survey it is evident that S-layers represent patterning structures at the ultimate resolution limit for a macromolecular functionalization of surfaces. The great potential of S-layers in molecular nanotechnology was clearly recognized when the capability for recrystallization of isolated subunits on solid surfaces and liquid interfaces was demonstrated.[14-17] In the following a broad spectrum of applications will be defined ranging from S-layers as nanonatural resists to matrices and patterning structures for the controlled immobilization of functional molecules. Particularly the observation that S-layer subunits have the capability to recrystallize at lipid layers (Langmuir-Blodgett films) led to a new concept of biomimetic membranes which copies the structural and functional principle of those archeaobacterial cell envelopes which are exclusively composed of a plasma membrane and a closely associated S-layer (see also chapter 2). Technologies based on S-layer stabilized artificial or biological lipid membranes should allow for the first time to exploit specific membrane functions in a macroscopic scale.

8.2. FORMATION OF S-LAYER LATTICES ON SOLID SUBSTRATES AND LIQUID SURFACE INTERFACES

8.2.1. INTERMOLECULAR FORCES

In most S-layer lattices the constituent subunits interact with each other and with the supporting cell envelope layer (e.g. plasma membrane, outer membrane or peptidoglycan) by a combination

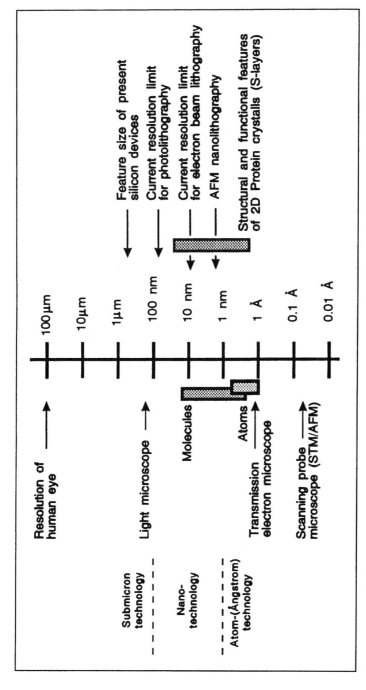

Fig. 8.1. Size scale for micro- and nanostructures. The left side summarizes the resolution capabilities of light-, electron-, and scanning probe microscopes while the right side shows the milestones in microfabrication technology with respect to S-layers lattices. The size of the constituent S-layer protein or glycoprotein subunits lies in the nanometer range. The center-to-center spacings between the morphological units are in the range of 3 to 30 nm. Since S-layers are composed of identical subunits functional groups are arranged on the regular arrays in well defined positions and orientations even down to the subnanometer scale.

of non-covalent forces including hydrogen or ionic bonds, and hydrophobic or electrostatic interactions.[18-27] Since S-layer proteins are composed of a high proportion of non-polar amino acids, hydrophobic interactions are expected to play a particularly important role in the assembly process.[28] Some S-layers have been shown to be stabilized by divalent cations interacting with polar groups. In particular, studies on the distribution and functional significance of charged groups on S-layer lattices of *Bacillaceae* have shown that free amino and carboxylic acid groups of adjacent protomers are arranged in close proximity and thus contribute to the stability of the array by electrostatic interactions. Isolation and disintegration experiments with S-layer lattices clearly demonstrated that the intermolecular forces between the subunits are stronger than those binding the crystalline array to the supporting envelope layer. This property is seen as a basic requirement for the dynamic recrystallization and reorganization process of the S-layer lattice on the bacterial cell surface in the course of cell growth and cell division (see also chapter 2).[29-32]

8.2.2. SELF-ASSEMBLY IN SUSPENSION

S-layers isolated from a broad spectrum of prokaryotic organism have shown the inherent ability to reassemble into two-dimensional arrays after removal of the disrupting agent used in the dissolution procedure.[31,33,34] Studies on this self-assembly process have shown that the initial phase is determined by a rapid nucleation of the subunits into oligomeric precursors consisting of several unit cells. Subsequently, these aggregates reassemble in a much slower process into larger crystalline arrays[28] (Fig. 8.2). Studies on the self-assembly process of S-layer subunits have shown that the formation of the crystalline arrays is an entropy driven process and only determined by the amino acid sequence of the polypeptide chains and consequently the tertiary structure of the S-layer protein species. The self-assembly products have the form of flat sheets, open ended cylinders or closed vesicles.[18,19,31,33,35] Structural form and size of the self-assembly products depend on several environmental parameters such as temperature, pH, ion composition and/or ionic strength. S-layer self-assembly products from *Bacillus Thermoanaerobacter* or *Clostridium spp.* reach a size of several square micrometers (Fig. 8.3.a-b). Morphology and bonding properties of the S-layer subunits also determine the assembly route, the final reassembly form, and the possibility of forming mono- or double layers. In the case of double layered structures, the two constituent monolayers face each other either with their inner or their outer side.[19] Spiral growth initiates when a screw dislocation is formed in a very early stage of the assembly process.

8.2.3. CRYSTAL GROWTH AT INTERFACES

In order to understand the recrystallization of isolated S-layer subunits at interfaces and solid surfaces it is necessary to consider

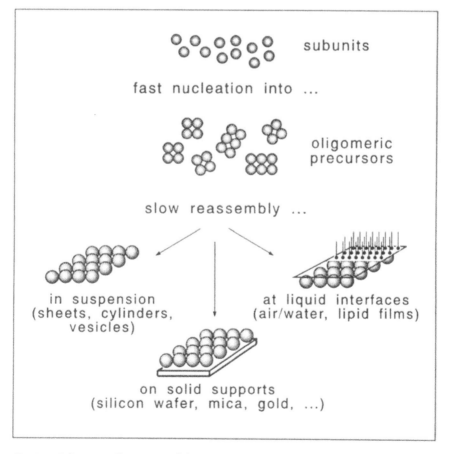

Fig. 8.2. Schematic illustration of the two-stage process of recrystallization of S-layer subunits into crystalline arrays. The self assembly process can occur in suspension, on solid supports, at liquid interfaces or Langmuir-Blodgett films.

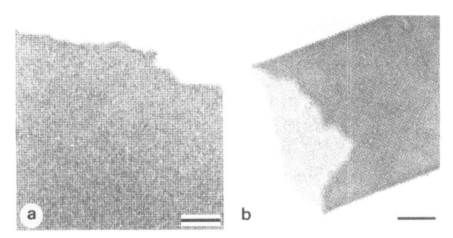

Fig. 8.3. Electron microscopical images of negatively stained preparations of S-layer self assembly products from Bacillus stearothermophilus NRS2004/3a/v2 obtained upon recrystallization of isolated subunits in suspension. (a) Flat sheet and (b) cylindrical self-assembly products. The S-layer shows square (p4) lattice symmetry with center-to-center spacing between the morphological units of 13.5 nm. Bars = 200 nm.

their surface properties. Structural and chemical analysis and adsorption studies have shown that S-layer lattices are highly anisotropic structures with regard to their inner and outer surfaces. Most detailed studies have been performed with S-layer lattices from *Bacillaceae*.[14,19,20,25,36-42] Generally the outer surface is more hydrophobic than the inner one. The inner surface reveals a net negative charge due to an excess of carboxylic acid groups whereas the outer surface is charge neutral due to an equimolar amount of carboxylic acid and amino groups. A further characteristic feature of most S-layers is their difference in surface topography. Three-dimensional computer image reconstructions of the protein mass distribution (for reviews see refs. 27, 43-45), ultrathin sections and freeze-dried preparations of S-layer self assembly products have shown that the outer surface is generally less corrugated than the inner one.

The orientation of the protein arrays recrystallized at interfaces is determined by the anisotropy of the physicochemical surface properties of S-layers lattices. Electron microscopical examinations revealed that recrystallized S-layers were oriented with their charge neutral, more hydrophobic outer face against hydrophobic solid surfaces, the air/water interface and with their negatively charged,

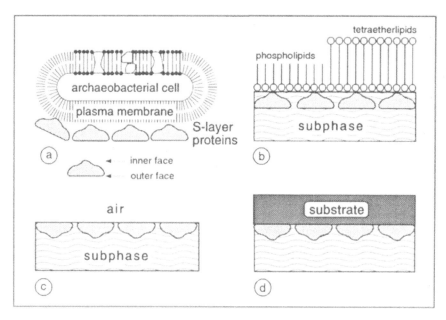

Fig. 8.4. Schematice illustration of the lattice orientation of an S-layer: (a) on an intact archeaobacterial cell possessing an envelope composed exclusively of an S-layer and a closely associated plasma membrane, (b) on phospholipid and tetraetherlipid films spread by Langmuir-Blodgett techniques, (c) at the air/water interface, and (d) on a hydrophobic solid surface.

more hydrophilic inner face against hydrophilic charge neutral, charged or zwitter-ionic headgroups of phospholipid films[14-17] (Fig. 8.4). The orientation of a recrystallized S-layer lattice can be determined by electron microscopical procedures and scanning force microscopy. This is particularly easy with S-layer lattices showing oblique lattice symmetry while higher spacegroup symmetries require careful analysis by image processing in order to determine the lattice orientation from the handedness of the morphological unit.[14-16]

Electron microscopical studies have shown that the recrystallization process at different interfaces including the air/water interface, lipid monolayers generated by LB techniques and solid supports is initiated at several distant nucleation points and advances in-plane until neighboring, also, growing crystalline areas are met[15,16] (Fig. 8.5). The nucleation sites may be either protomers, oligomers or small self-assembly products adsorbed from the solution. Since the lateral orientation of the nucleation points

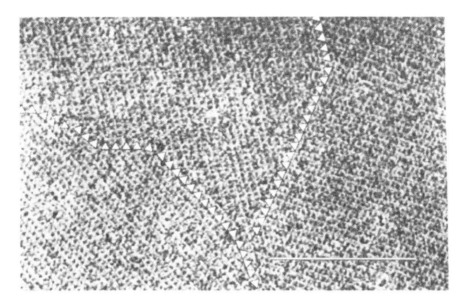

Fig. 8.5. Electron micrograph illustrating the oblique (p1) S-layer lattice of Bacillus coagulans E38-66/v1 recrystallized at the air/water interface. Center-to-center spacings between the morphological units are 9.4 nm and 7.4 nm, respectively, and the base angle is 80°. The crystal boundaries are seen as line imperfections between the crystalline domains (marked by arrows). Bar = 200 nm.

determines the orientation of the crystalline domains a closed mosaic of monocrystalline areas is obtained. The average size of the monocrystalline domains depends on the density of the nucleation sites and the lateral mobility of the protein subunits in close vicinity to the incorporation sites. Depending on the bacterial strain from which the S-layer was derived and the substrate used an average patch size of up to 15 μm in diameter may be obtained.[16] Crystals formed in the described way grow under equilibrium conditions. Fascinating structures are obtained when crystal growth takes place under non-equilibrium conditions.[46,47] For the recrystallization of the S-layer of *Bacillus sphaericus* CCM 2177 the calcium concentration in the subphase was shown to be of significant importance. Depending on this parameter a broad spectrum of crystal morphologies ranging from tenuous, fractal-like structures to large monocrystalline patches was found.[17] Figure 8.6 shows an example for a perfect crystalline and a tenuous S-layer structure of this strain. Although all these structures look like fractals obtained by diffusion limited aggregation, they are not aggregates of randomly oriented protein subunits. Image processing methods have shown that all morphological units in the cluster follow the lattice orientation of the crystal lattice.[17]

8.2.4. RECRYSTALLIZATION ON SOLID SURFACES

S-layer proteins isolated from numerous organisms have shown the ability to recrystallize on solid surfaces such as silicon, mica, carbon or synthetic polymers.[8-13] The environmental conditions required for the crystallization process are frequently the same as those used to induce self-assembly without supporting layers. The clear supernatant (containing ~1 to 2mg/ml S-layer protein) is injected into a glass dish filled with buffer. After thorough mixing the solid substrate is put immediately onto the liquid surface interface. Recrystallization conditions and time have to be optimized for each S-layer individually. The time depends not only on the type of S-layer and the particular substrate but also on environmental conditions such as temperature, pH or ion composition and/or ionic strength of the subphase. Generally, the formation of a coherent monolayer on the substrate takes a few hours. Finally, the S-layer coated substrate is carefully removed from the surface by horizontal lifting.

In general as a precondition for many S-layer assembly systems the surface of the supporting layer has to be rendered hydrophobic.

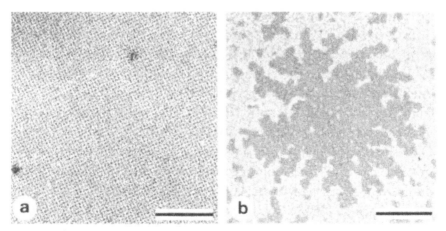

Fig. 8.6. Electron micrographs of negatively stained preparations of the square (p4) S-layer of Bacillus sphaericus CCM2177 obtained upon recrystallization at the air/water interface. (a) Perfect crystalline S-layer. Bar = 250 nm. (b) Corresponding tenuous S-layer structure obtained at lower calcium concentrations in the subphase. The center-to-center spacing of the morphological units is 14.5 nm. Bar = 500 nm.

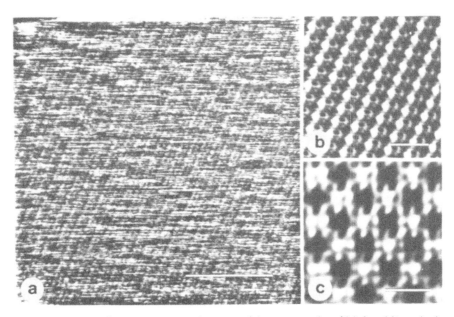

Fig. 8.7. Scanning force microscopical images of the topography of (a) the oblique (p1) S-layer lattice of B.coagulans E38-66/v1 (Center-to-center spacings of the morphological units are 9.4 nm and 7.4 nm, base angle 80°; Bar = 100 nm), (b) the corresponding computer image reconstruction (Bar = 20 nm), and (c) of the square S-layer of Bacillus sphaericus CCM2177 (Center-to-center spacing of the morphological units 14.5 nm, computer image reconstruction, Bar = 20 nm). In all images Black-to-white corresponds to a corrugation of 2 nm.

S-layer subunits from several *Bacillaceae* have shown to adsorb with their more hydrophobic outer face on charge neutral hydrophobic surfaces.[14] Silanization procedures with different compounds such as octadecyltrichlorosilane (OTS) or 3-(trimethoxysilyl)propyl-methacrylate (TSP) can be used to produce hydrophobic surfaces on silicon or glass. S-layers recrystallized on solid supports are best studied by scanning force microscopy. Figure 8.7 shows scanning force images of two different S-layers. Both recrystallized on an OTS-silanized silicon wafer. The images were recorded under fluid in contact mode. The nominal loading force for obtaining high resolution images had to be in the range of 100 pN. Increasing the force up to 1 nN led to a loss in resolution and in the following to complete destruction of the native S-layer.

S-layers which have recrystallized on solid surfaces can be used for a variety of applications in nanomanufactoring.[8-13,48-50] For this purpose supports have to be selected which fulfill specific requirements such as stiffness or flatness. Further on, the availability of functional groups on the surface can be important if a specific orientation of the recrystallized S-layer lattice or crosslinking of the protein array to the supporting layer is required. Flatness and mechanical properties are particularly important in cases where a scanning force microscope will be used as a molecular assembler for manipulating elements at molecular level.[51,52]

Materials commonly used in nanomanufacturing are silicon and precious metals. Silicon has proven to be one of the most potent materials for the fabrication of (sub)micrometer structures and machines (for review see ref. 53). It is atomically flat, chemically well characterized and may be produced with almost perfect crystallinity. Precious metals such as platinum or gold which are chemically inert are also important since they allow the investigation of redoxreactions in molecular assemblies by electrochemical measurements. Table 8.1 summarizes solid substrates which have

Table 8.1. List of solid supports successfully used in S-layer recrystallization experiments

Material	Surface modification
Polished silicon wafer	(a) None (as supplied by the manufactorer) (b) Cleaned in hot basic and acidic hydrogenperoxide (c) Silanized with octadecyltrichlorosilane (OTS), or (d) 3-(trimethoxysilyl)propylmethacrylate (TSP) (e) Spin-coated with a photoresist (AZ1530, Hoechst)
Mica	Silanized with OTS
Highly oriented pyrolytic graphite (HOPG)	None
Glass	Formvar/Carbon coated
Electron microscopical grids	(a) Formvar/carbon coated (b) Formvar/Poly-L-lysine coated
Gold	Functionalized with different SH-lipids

been already tested for their suitability as matrix for recrystallizing S-layer lattices as coherent layers.

8.2.5. REASSEMBLY OF S-LAYER PROTEINS AT THE AIR/WATER INTERFACE AND ON LIPID FILMS

Reassembly of isolated S-layer subunits at the air/water interface and on lipid films has proven to be an easy and reproducible way for generating coherent S-layer lattices at large scale.[15,16] In accordance with S-layers recrystallized on solid surfaces the orientation of the protein arrays at liquid interfaces is also determined by the anisotropy in the physicochemical surface properties. Electron microscopical examinations revealed that recrystallized S-layers were oriented with their charge neutral, more hydrophobic face against the air/water interface and with their negatively charged, more hydrophilic inner face against the hydrophilic charge neutral, charged or zwitter-ionic headgroups of phospho- or tetraether lipid films. Table 8.2 summarizes various lipids which have already been successfully used in the recrystallization experiments. Figure 8.8. shows an electron micrograph of a negatively stained

Table 8.2. List of lipids and fatty acids successfully used in S-layer recrystallization experiments

Short name	Full name
DPPC	Dipalmitoylphosphatidylcholine
DPPE	Dipalmitoylphosphatidylethanolamine
DMPE	Dimyristoylphosphatidylethanolamine
DSPE	Distearoylphosphatidylethanolamine
DPPA	Dipalmitoylphosphatidic acid
DOPE	Dioleoylphosphatidylethanolamine
DLPE	Dilinoleoylphosphatidylethanolamine
DHPC	Dihexadecylphosphatidylethanolamine
AraSre	Arachidic acid
SteaSre	Stearic acid
Dilysin	Nε-Lysyl-Nε-lysin-octadecylamide
GDNT	Glyceroldialkylnonitoltetraether
DPPC=	Dipalmitoleoylphosphatidylcholine
SOPC	Stearoyl-Oleoyl-phosphatidylcholine
POPC	Palmitoyl-Oleoyl-phosphatidylcholine
HDA	Hexadecylamine
Ch	Cholesterol

Fig. 8.8. Electron micrograph of a negatively stained composite S-layer/lipid film transferred by horizontal lifting onto an electron microscopical grid. The S-layer of Bacillus coagulans E38-66/v1 shows oblique (p1) lattice symmetry with center-to-center spacings between the morphological units of 9.4 nm and 7.4 nm, respectively, and a base angle of 80°. The dipalmitoylphosphatidylethanolamine(DPPE)-film was spread by LB-techniques. The S-layer subunits were subsequently recrystallized from the subphase. (See also Fig. 8.4.b) Bar = 100 nm.

composite S-layer/lipid film transferred by horizontal lifting onto an electron microscopical grid.

Lipid films are formed at the air/water interface by standard LB-techniques (for review see refs. 54 and 55). Lipids dissolved in organic solvents are spread on the surface and compressed to a certain surface pressure after evaporation of the solvent. Depending on the surface pressure the lipid molecules are either in an expanded, partly compressed or close packed phase. Experimental data indicated that particularly the fluidity of the lipid film is a very critical parameter for generating a coherent S-layer.[56] The fluidity of the film may also be visualized directly by Brewster-angle microscopy.[57] Both fluorescence and Brewster-angle microscopy have shown that the S-layer subunits frequently start to recrystallize at the borderline of lipid islands (packed phase) and continues under the expanded phase. Figure 8.9 shows a phase separated DPPE-film imaged by Brewster-angle microscopy. The growing S-layer is seen as an irregular borderline.

The mechanical stability of S-layers and composite S-layer/lipid films at interfaces can be increased by inter- and intramolecular crosslinking of the subunits from the subphase. This stabilization procedure has been shown to prevent particularly destruction of

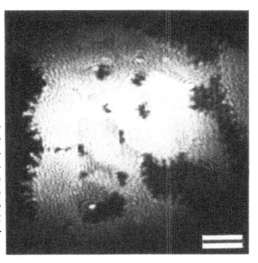

Fig. 8.9. Phase separated dipalmitoylphosphatidylethanolamine (DPPE)-film imaged by Brewster-angle microscopy. The recrystallizing growing S-layer is seen as the irregular boundary line between crystalline and liquid phase of the lipid film. The image was taken 5 minutes after injection of the S-layer proteins into the subphase. Bar = 50 µm.

the S-layer or S-layer/lipid membranes during subsequent handling procedures. For example, covalent linkages can be introduced with glutaraldehyde which acts as a zero-length crosslinker between amino groups. The proportion of lipid molecules in the monolayer that can be covalently linked to the porous S-layer lattice will significantly modulate the lateral diffusion of the free lipid molecules and consequently the fluidity of the whole membrane. We have introduced the terminology "semifluid membrane" for such composite structures[16] (Fig. 8.10). A certain degree of inhibition of fluidity in the lipid film associated with the S-layer lattice can be expected even without crosslinking. Lipid head groups in monolayer films associated with an S-layer will interact specifically with defined domains of the protein lattice. S-layers which have been recrystallized at liquid surface interfaces or lipid films can be transferred onto solid supports by standard LB-techniques such as vertical or horizontal deposition, or by surface lowering (Fig. 8.11). Depending on the type of lipid (phospholipid, tetraetherlipid) and the transfer procedure applied, different composite layers can be generated. For examination by electron microscopical procedures recrystallized S-layers and composite S-layer/lipid films can be transferred onto standard or perforated Formvar/carbon coated electron microscope grid.[58,59] The grids were carefully placed onto the interface and removed from it after several seconds by hand with

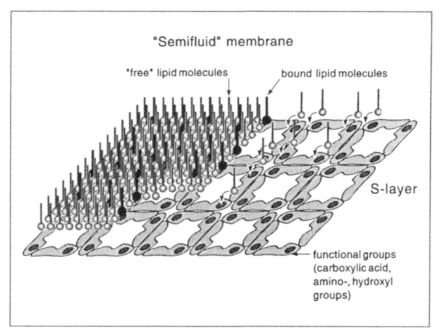

Fig. 8.10. Schematic illustration of the "semifluid membrane model" (see text for details).

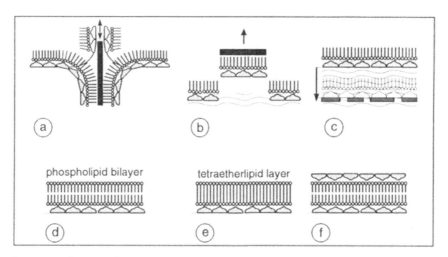

Fig. 8.11. Schematic illustration of transferring an S-layer or composite S-layer/lipid film onto a solid or porous support. (a) Vertical, (b) horizontal lifting (Schaefer method), and (c) surface lowering. (d,e,f) Depending on the type of lipid (phospholipid, tetraetherlipid) and the transfer procedure applied, different composite layers can be generated.

forceps. After negative staining (e.g. uranyl-acetate) the protein and protein/lipid monolayers could be clearly imaged in the electron microscope. Special care had to be taken with "holey" grids because non-supported S-layers and S-layer/lipidfilms covering the holes were only stable in the electron beam when minimum dose conditions were used. It was found that S-layer stabilized lipid films are able to span holes with a diameter of up to 15 μm.[16] As discussed later, these composite structures which mimic the molecular architecture of those archaeobacterial cell envelopes that are exclusively composed of an S-layer and a plasma membrane could lead to new techniques for exploiting large scale structural and functional principles of membrane associated and integrated molecules (e.g. ion channels, proton pumps, receptors).

8.3. S-LAYERS AS PATTERNING STRUCTURES AND NANONATURAL RESISTS IN MOLECULAR NANOTECHNOLOGY

8.3.1. CONTROLLED IMMOBILIZATION OF FUNCTIONAL MOLECULES

The controlled immobilization of molecules on surfaces is an essential requirement in most areas of supramolecular engineering and molecular nanotechnology. Contrary to conventional carriers where the location, local density and orientation of functional groups and the porosity and pore size are only known approximately, with S-layers lattices, the properties of a single constituent unit are replicated with the periodicity of the lattice and thus define the characteristics of the whole two-dimensional array. According to this principle it has already been shown that macromolecules may be bound to S-layers in dense crystalline packing[8,60-62] (see also chapter 6). Specific binding of molecules on S-layer lattices may be induced by different non-covalent forces. The pattern of binding frequently reflects the lattice type, the size of the morphological units and the distribution of physicochemical properties on the array. For example, the distribution of net negatively charged sites on S-layers can be visualized by electron microscopical methods after labeling with positively charged topographical markers such as polycationic ferritin (PCF; diameter, 12 nm). The regular arrangement of free carboxylic acid groups on the hexagonal S-layer lattice from the archeaobacterium *Thermoproteus tenax* could be clearly demonstrated in this way.[63] Furthermore, S-layers have also shown to be particularly suitable as a matrix for a covalent attachment of functional macromolecules.[8,60-62] For this purpose, carboxylic acid groups on the S-layer protein can be activated with carbodiimide while hydroxyl groups of the carbohydrate chains of S-layer glycoproteins are generally treated with cyanogen bromide or periodate. So far S-layer lattices have been used as immobilization matrix for a broad spectrum of macromolecules of biologically active proteins (e.g. enzymes, antibodies, ligands) (see also chapter 6).

Based on these principles, a broad range of amperometric and optical bioanalytical sensors was developed.[64-67] In these devices enzymes were covalently linked to S-layer fragments or S-layer self-assembly products. For the fabrication of a single enzyme sensor, such as a glucose sensor, glucose oxidase (GOD) molecules are covalently bound to the exposed S-layer surface of an S-layer ultrafiltration membrane[64] (Fig. 8.12a) (see also chapter 6). The electrical contact to the sensing layer was established by sputtering a thin layer of platinum or gold onto the enzyme layer (Fig. 8.12b). The whole assembly was pressed against a solid gold plate in order to increase the mechanical stability of the membrane (Fig. 8.12c). The analyte reaches the sensing layer through the open structure of the microfilter. In the course of the enzymatic reaction of glucose with glucose oxidase, gluconic acid and hydrogen peroxide are produced under consumption of oxygen. The glucose concentration is determined by measuring the current of the electrochemical oxidation of hydrogen peroxide. The response curve of an S-layer glucose sensor is shown in Figure 8.12d. For the construction of multi-enzyme sensors a different construction principle was developed since the simultaneous immobilization of different enzyme species generally leads to an uncontrollable competition for the available activated binding sites.[65] For producing multienzyme biosensors, individual enzymes are first immobilized to S-layer fragments. Subsequently the different enzyme loaded S-layer fragments are mixed in suspension, deposited on a microfiltration membrane and sputter coated with platinum or gold. In this way immobilization parameters could be optimized for each enzyme individually

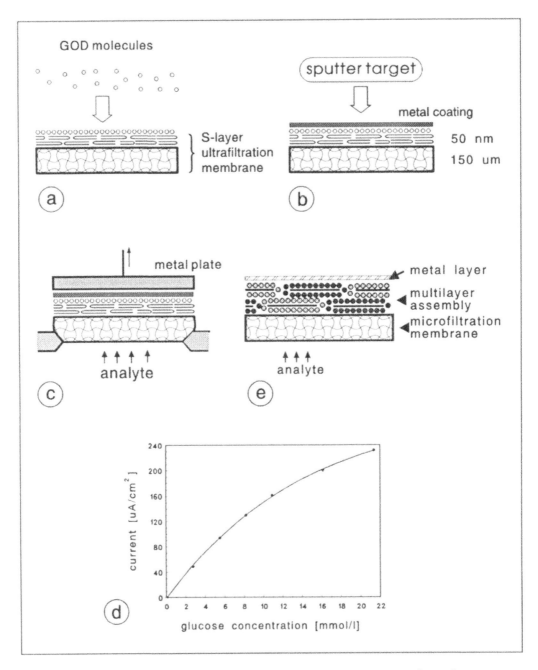

Fig. 8.12. (a to b) Schematic illustration of the fabrication of an amperometric S-layer glucose sensor. (a) The sensing layer consists of an S-layer ultrafiltration membrane onto which a dense monolayer of enzyme molecules (e.g., glucose oxidase, GOD) has been covalently bound. (b) The electrical contact to the enzyme molecules is established by depositing a platinum or gold layer on the sensing layer by sputter coating. (c) The analyte reaches the sensing layer through the open structure of the microfilter. A metal plate is used for supporting the sensing layer. (d) Response curve of an S-layer glucose sensor. (e) Schematic illustration of the construction principle of a multienzyme S-layer biosensor.

and the ratio of the amounts of enzyme molecules accurately controlled (Fig. 8.12e). This method leads to a well structured sandwich of thin monomolecular enzyme layers where protective layers can easily be integrated. On the basis of this technique several multienzyme sensors such as a sucrose sensor with three enzymes species (invertase-mutarotase-glucose oxidase) or a cholesterol sensor (with cholesterol esterase and cholesterol oxidase) were developed. The various types of amperometric S-layer biosensors are summarized in Table 8.3.

Furthermore, such S-layer based sensing layers were also used in the development of optical biosensors where the electrochemical transduction principle was replaced by an optical one (Fig. 8.13a). In this approach a pH- or oxygen-sensitive fluorescent dye was immobilized on the S-layer in close proximity to the glucose oxidase sensing layer. In the pH-sensitive system, with carboxyfluorescine as pH-sensitive dye, the glucose concentration in the sample was derived from the change in pH in the microenvironment between the dye and the enzyme molecules. The pH-drop

Table 8.3. Types of single- and multienzyme amperometric biosensors based on S-layer technology

Type of sensor	Enzymes	Signal ($\mu A*l/cm^{2*}mmol$)	Linear range	Application
glucose	glucose oxidase	15	0–20 mmol glucose	blood analysis, microbiology
ethanol	alcohol oxidase	2.5	0–7 mmol ethanol	blood analysis, microbiology, food technology
xanthine	xanthine oxidase	30	0–0.6 mmol xanthine	food technology
maltose	maltase + glucose oxidase	1.5	0–1.5 mmol maltose	microbiology
sucrose	β-fructosidase + mutarotse + glucose oxidase	1	0–12 mmol sucrose	microbiology, food technology
cholesterol	cholesterol esterase+ cholesterol oxidase	6	0–0.5 mmol cholesteryl palmitate	blood analysis

itself was caused by the enzymatic reaction due to the production of gluconic acid. For the oxygen sensitive system a Ruthenium(II)-complex was used as transducer. The fluorescence of the Ru(II)-complex is dynamically quenched by molecular oxygen. Thus, a decrease in the local oxygen pressure as a result of the enzymatic reaction can be detected and transformed into a measurable signal. In an alternative approach the design of the oxygen

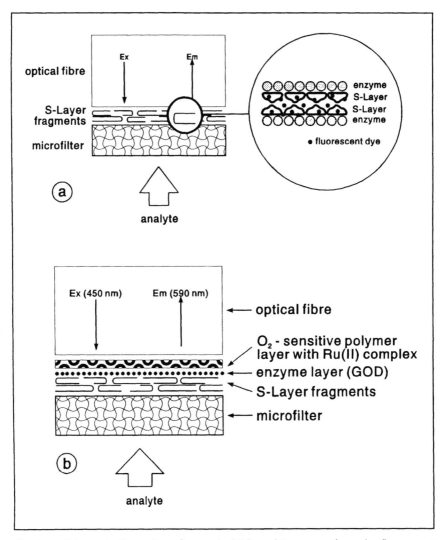

Fig. 8.13. Schematic illustration of an optical S-layer biosensor where the fluorescent dyes are (a) immobilized in close proximity to the enzyme loaded sensing layer, or (b) bound in a seperate thin polymeric matrix (see text for details).

sensitive S-layer optode was modified by separating the biological sensing layer from the optical sensing element[66] (Fig. 8.13b). The biological sensing layer was made of an enzyme loaded S-layer ultrafiltration membrane which had already been optimized in the development of amperometric biosensors. The fluorescent dyes were bound in a thin foil of a polymeric matrix. This design allowed optimization the components individually. Furthermore, a broad range of biologically sensitive layers (see Table 8.3) could be used with the same optical system. S-layers as immobilization structures for biologically active molecules were also used in the development of an infrared optical biosensor. In this device an S-layer which had been recrystallized on the cylindrical part of an infrared transparent optical fiber (chalkogenide fiber) was used to bind glucose oxidase. The glucose concentration was determined from the infrared spectrum of gluconic acid.[67]

8.3.2. WRITING WITH MOLECULES

The scanning force microscope (SFM) may not only be used to image biological structures at molecular resolution (for several examples on scanning force microscopy on two-dimensional crystalline biological structures see refs. 68-76). It has also been introduced as a "molecular assembler". SFM is currently the only tool which is fine enough to handle individual atoms and molecules on plane surfaces.[77-79] An essential requirement for assembling molecular devices composed of biologically active molecules on the nanometer level is the availability of a geometrically and physicochemically precisely defined regular immobilization matrix which resembles in its pattern scale the size of the respective molecules. So far, the application of S-layers recrystallized on solid supports as a matrix for immobilizing biomolecules clearly demonstrated that the protein lattices are perfectly suited patterning structures on the molecular level (see also chapter 6). Corresponding to the classic work of Eigler and coworkers[77,78] who have demonstrated the possibility of writing with atoms on solid metal surfaces, it should be possible to specifically manipulate macromolecules with the SFM-tip on the surface of S-layer lattices under aqueous environments (Fig. 8.14). To prove this concept we suggested to use as a model system polycationic ferritin (PCF) electrostatically bound in well

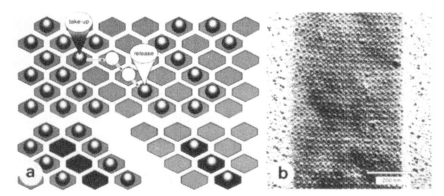

Fig. 8.14. Schematic illustration of writing with molecules on a hexagonal S-layer lattice. (a) Molecules which are bound by electrostatic forces onto the S-layer may be transferred with the scanning force microscope (SFM)-tip to a blank morphological unit in a controlled way. Depending on the prospective application a negative (bottom left) or positive (bottom right) writing process can be applied. (b) Polycationized ferritin (PCF) molecules are immobilized in a regular fashion by electrostatic interactions on the hexagonal S-layer of Thermoproteus tenax. *The center-to-center spacing between the morphological units of the S-layer lattice and the PCF molecules is 30 nm. Bar = 200nm.*

defined positions onto negatively charged domains on an S-layer lattice with large lattice spacings.[9,12,13] The bound PCF molecules could be dragged with an electrically negatively charged (metal coated) SFM-tip and dropped at the negatively charged domains of the S-layer lattice, by reversing the polarity of the tip. Due to the crystalline arrangement of the negatively charged domains on the S-layer lattice the deposited PCF-molecules could generate geometric patterns resembling the periodicity of the underlying lattice. Theoretically "writing with molecules" on S-layer lattices could also be done by selectively removing individual molecules from a lattice which in an initial step had been completely loaded with the respective molecules. Reading the information would only require scanning the native structure or a high resolution replica of it with the SFM-tip again.[80,81] Theoretically, the storage capacity of such a device would be one terabit of binary information on 1 cm² for a square S-layer lattice with a lattice spacing of 10 nm. Currently, the most severe problems of such a new storage technology are seen in the necessity to store or read the information sequentially, in a quick erasure of the information and very critical environmental parameters for maintaining the integrity of a "biological storage device" over years. On the other hand if high resolution replicas of such structures are used for permanent data storage this problem will not be relevant.

8.3.3. S-LAYERS AS NANONATURAL RESIST

Another approach for exploiting the potential of S-layer lattices in nanotechnology is based on making submicron structures on S-layer coated silicon wafers by exposure with deep ultraviolet radiation (DUV; 193 nm). This technique has already been used for "at-the-surface" imaging in the field of microfabrication.[48-50,82-86] In one approach a thin refractory layer (Zr, deposited via reaction with $ZrOCl_2$) was deposited on an organic resist surface and amplified by a multilayer self-assembly process.[85] In this way an etching mask was formed for subsequent plasma etching. In another approach the binding ability of the organic layer was modified at areas which previously had been exposed to the radiation. The remaining unexposed areas were then either primed to increase the number of binding sites and selectively coated with refractory clusters,[85] or directly used for binding active biomolecules.[84] Preliminary studies with S-layers have shown that the S-layer is removed specifically from silicon wafers by DUV but retained its crystalline and functional integrity in the unexposed areas. In a similar approach to the work described for nanonatural resists, unexposed S-layer areas could be used either to bind enhancing ligands or to enable electrodeless metallization (Fig. 8.15). In both cases a layer is formed which allows a patterning process by oxygen reactive ion etching. Alternatively unexposed S-layers may also be used for selectively binding lipid layers and/or biologically active molecules which would be necessary for the development of supramolecular structures. Since S-layers are only 5 to 10 nm thick and consequently much thinner than conventional resists, considerable improvement in edge resolution in the fabrication of submicron structures can be expected.

8.4. BIOMIMETIC APPLICATIONS

The possibility for recrystallizing isolated S-layer subunits into large scale isoporous, coherent protein lattices at an air/water interface or on lipid films and for handling such layers by standard Langmuir-Blodgett techniques[15,16] opens a broad spectrum of applications in basic and applied membrane research including physiology, diagnostics and biosensor developments. These composite structures strongly resemble those archaeobacterial envelope structures which are exclusively composed of an S-layer and a closely associated plasma membrane[24,87-89] (see also chapter 2). Since many of these organism dwell under extreme environmental

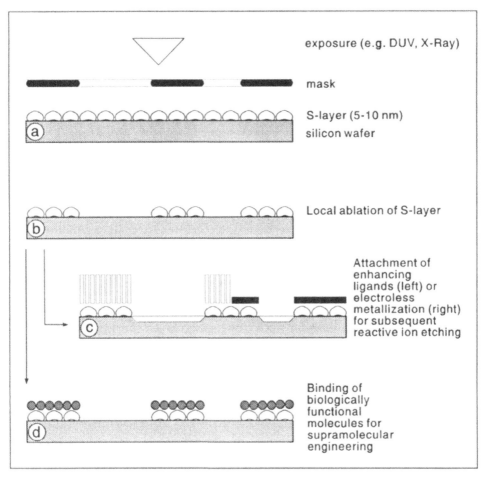

Fig. 8.15. At-the-surface imaging with S-layers as nanonatural resist. (a) A pattern is transferred onto the S-layer by exposure (e.g. deep ultraviolett radiation (DUV)) through a microlithographic mask. (b) The S-layer is removed specifically from silicon wafers but retains its crystalline and functional integrity in the unexposed areas. (c) Unexposed S-layer areas could be used either to bind enhancing ligands or to enable electrodeless metallization. In both cases a layer is formed which allows a patterning process by oxygen reactive ion etching. (d) Alternatively unexposed S-layers may also be used for selectively binding lipid layers and/or biologically active molecules which would be necessary for the development of supramolecular structures.

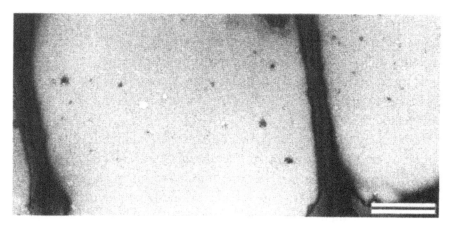

Fig. 8.16. Electron micrograph of a holey grid completely covered with a monomolecular S-layer recrystallized on a dipalmitoylphosphatidylethanolamine (DPPE) monolayer film after crosslinking the layers from the subphase with glutaraldehyde. (S-layer of Bacillus sphaericus CCM2177, square (p4) lattice symmetry, center-to-center spacing of the morphological units 14.5 nm) Bar = 500 nm.

conditions (e.g. high temperatures, low pH, high salt concentrations), the S-layers must have a strong stabilizing effect on lipid membranes. We have shown that S-layer supported LB-films and biological membranes (Fig. 8.16) can cover holes or apertures up to several microns in diameter[16] and maintain their structural and functional integrity in the course of subsequent handling procedures for a much longer period of time in comparison to unsupported structures (e.g. black lipid membranes). The stabilizing function of S-layers is primarily explained by a reduction or inhibition of horizontal vibrations which are seen as main cause for disintegration of planar lipid membranes. Particularly stable composite structures will be obtained after intra- and intermolecular crosslinking the S-layer proteins alone or with molecules from the lipid layer from the subphase (e.g. with glutaraldehyde). Additional lipid layers or S-layer supported lipid layers can be deposited on such "semifluid membranes" by standard LB-techniques (Fig. 8.11) or by fusion of lipid vesicles. Functional molecules may be incorporated into S-layer stabilized lipid layers using well established procedures. The most promising candidates for such studies are functional molecules such as carriers, ion channels, proton pumps, photo reaction centers, light harvesting and receptor molecules. Using single aperture silicon discs or patch-clamp pipettes, transport phenomena will only be determined by those molecules which are located over the opening (Fig. 8.17). On such stabilized lipid membranes functional measurements (electron or ion transport phenomena, binding of molecules to receptors, etc.) may be complemented by high resolution electron and scanning probe microscopy.

Presently there are barely any alternative biomimetic approaches other than S-layer supported membranes available for exploiting specific functional principles of lipid membranes at macroscopic scale. Our long-term strategy focuses primarily on the use of S-layer ultrafiltration membranes (see also chapter 6) as large scale support

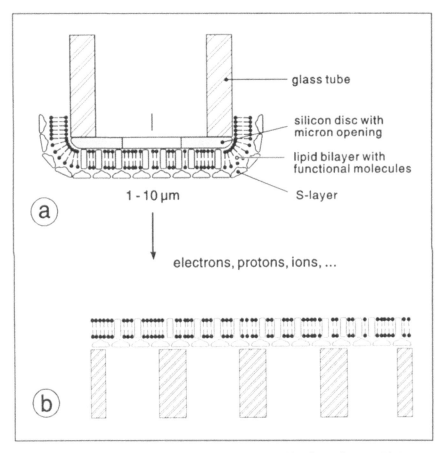

Fig. 8.17. Schematic illustration of an S-layer supported lipid membrane with incorporated functional molecules spanned over (a) the aperture of a silicon disk, a patch-pipette or (b) a porous support.

for functional lipid mono- or double layers (e.g. phospholipids or tetraetherlipids). Preliminary studies have shown that mono-, double or multilayers of surfactants generated by LB-techniques can be even directly transferred onto the surface of S-layer ulrafiltration membranes (Heckmann, personal communication). Such composite structures function as solution diffusion membranes at very low pressure gradients. An obvious area for application for this membranes will be in desalination processes[90-94]

We are not only concentrating our investigations on planar biomembranes but also on liposomes coated with S-layers as immobilization structures for macromolecules.[95] In these studies isolated S-layer subunits from different Bacillus species were recrystallized on positively charged liposomes. The S-layer subunits were shown to attach to the liposomes with their negatively charged inner face in an orientation identical to the lattice on the bacterial cell. The S-layer protein once recrystallized on liposomes was cross-linked with glutaraldehyde and subsequently used as a matrix for the covalent attachment of macromolecules (see also chapter 6). Based on these results we have started to investigate S-layer coated liposomes with incorporated purple membrane fragments or isolated bacterial rhodopsin.[96,97] It could be shown that the function of the proton pump is not hampered in liposomes completely covered with an S-layer lattice (Therese Lackner, personal communication). The high mechanical stability of S-layer coated liposomes and the possibility for immobilizing or incorporating biologically active molecules in such biomimetic structures may have great impact on various different liposome applications.

8.5. CONCLUSION

The present application potential for S-layers as patterning structures in molecular nanotechnology, as immobilization matrices in supramolecular engineering and as biomimetic membranes is based on 20 years of basic research on the structure, genetics, chemistry, morphogenesis and function of S-layers. In the course of these investigations a broad spectrum of physical and biological methods and preparation techniques was used in an innovative way leading to an interdisciplinary research area at the interface between biology and physics. This approach led to the development of a technology base which allows the functionalization of surfaces with biological crystalline arrays and the fabrication of "new advanced materials". In this chapter, we have primarily focused on applications for macroscopic S-layer lattices which have already been developed up to the stage of working models. Nevertheless, there are many more ideas for S-layer based technologies which should be addressed briefly. For example, biomineralization is becoming increasingly important in the structural, interfacial and architectural design of inorganic materials.[98,99] Here are three main streams in biomimetic materials chemistry currently being developed within the context of biomineralization. First, specific chemical and structural properties of biomolecules can be used in synthetic reactions, second, use of living organisms such as bacterial cultures as intact biosystems to make highly ordered composites and third, biomaterial chemistry involving biomineralization as an inspiration for new ideas in the synthesis of inorganic materials exhibiting uniform particle size, morphology, oriented nucleation and assembly. S-layers may be used in this field as a geometrically precisely defined surface with accurately positioned potential nucleation sites for biomineralization processes. Recent studies on a cyanobacterial S-layer clearly demonstrated that a crystalline surface layer can be involved in the production of fine grain minerals in lakes.[100]

S-layers as isoporous structures may also be used for the fabrication of nanometric metallic point patterns.[50] In this approach a metal (e.g. silver) is thermally evaporated onto an S-layer recrystallized on a glass substrate. In the course of the metal deposition only those silver molecules penetrating the pores will form a periodic pattern of nucleation points on the substrate. After dissolution of the S-layer with appropriate chemicals the fabrication of a metal island film with an arrangement of islands according to the pore distribution in the S-layer lattice becomes possible. Such films have fascinating non-linear optical properties and are considered as very important in the development of a new generation of ultrafast optical computers.[101]

The unique features of S-layer lattices which have been optimized during billions of years of biological evolution will lead to many more applications than presented here. We also hope that this review will particularly stimulate "non-biologists" to enter the field of applied S-layer research.

ACKNOWLEDGMENTS

The assistance of Claudia Hödl, Therese Lackner, Angela Neubauer, Andrea Scheberl, Bernhard Schuster and Barbara Wetzer is gratefully acknowledged. Part of this work was supported by the Austrian Science Foundation (FWF) grants S7204 and S7205 and the Austrian Federal Ministry of Science, Research and the Arts.

REFERENCES
  1. Wilchek M, Bayer EA, eds. Avidin-biotin technology. Methods in enzymology. Vol. 184. Orlando: Academic Press, 1990.
  2. Wegner, G. Ultrathin films of polymers. Ber Bunsenges Phys Chem 1991; 95:1326-33.
  3. Whitesides GM, Mathias JP, Seto CT. Molecular self-assembly and nanochemistry: a chemical strategy for the synthesis of nano-structures. Science 1991; 254:1312-19.
  4. Nagayama K. Protein array: an emergent technology from biosystems. Nanobiology 1992; 1:25-37.
  5. Knoll W, Angermaier L, Batz G et al. Supramolecular engineering at functionalized surfaces. Synth Metals 1993; 61:5-11.
  6. Matsui S. Trends in nanostructure fabrication technology. FED J 1994; 4:34-43.
  7. Preece JA, Stoddart JF. Towards molecular and supramolecular devices. In: Welland ME, Gimzewski JK, eds. Ultimate limits of fabrication and measurement. Dordrecht: Kluwer Academic Publishers, 1995:1-8.
  8. Pum D, Sára M, Sleytr UB. Two-dimensional (glyco)protein crystals as patterning elements and immobilisation matrices for the development of biosensors. In: Sleytr UB, Messner P, Pum D, Sára M, eds. Immobilised macromolecules: application potential. London: Springer Verlag, 1992:141-60.
  9. Pum D, Sleytr UB. Molecular nanotechnology with S-layers. In: Beveridge TJ, Koval SF, eds. Advances in bacterial paracrystalline surface layers. New York: Plenum Press, 1992:205-18.
 10. Sleytr UB, Messner P, Pum D et al. Crystalline bacterial cell surface layers: general principles and application potential. J Appl Bacteriol 1992; 74:21S-32S.
 11. Sleytr UB, Sára M, Messner P, Pum D. Two-dimensional protein crystals (S-layers): fundamentals and applications. J Cell Biochem 1994; 56:171-6.
 12. Sleytr UB, Sára M, Messner P, et al. Application potentiol of 2D protein crystals (S-layers). In: Kelly RM, Wittrup KD, Karkare S, eds. Biochemical engineering VII. New York: Ann New York Acad Sci, 1994:261-9.

13. Pum D, Sára M, Sleytr UB. S-layers as molecular patterning structures. In: Welland ME, Gimzeski JK, eds. Ultimate limits of fabrication and measurement. Dordrecht: Kluwer Academic Publishers, 1995:197-203.

14. Pum D, Sára M, Sleytr UB. Structure, surface charge and self-assembly of the S-layer lattice from *Bacillus coagulans* E38-66. J Bacteriol 1989; 171:5296-303.

15. Pum D, Weinhandl M, Hödl C et al. Large scale recrystallization of the S-layer of *Bacillus coagulans* E38-66 at the air/water interface and on lipid films. J Bacteriol 1993; 175:2762-66.

16. Pum D, Sleytr UB. Large scale reconstitution of crystalline bacterial surface layer (S-layer) proteins at the air/water interface and on lipid films. Thin Solid Films 1994; 244:882-86.

17. Pum D, Sleytr UB. Anisotropic crystal growth of the S-layer of *Bacillus sphaericus* CCM 2177 at the air/water interface. Colloids and Surfaces A: Physicochem Eng Aspects 1995; 102:99-104.

18. Sleytr UB. Regular arrays of macromolecules on bacterial cell walls: structure, chemistry, assembly and function. Int Rev Cytol 1978; 53:1-64.

19. Messner P, Pum D, Sleytr UB. Characterization of the ultrastructure and the self-assemblies of the surface layer of *Bacillus stearothermophilus* strain NRS 2004/3a. J Ultrastruct Mol Struct Res 1986; 97:73-88.

20. Sára M, Sleytr UB. Charge distribution on the S-layer of *Bacillus stearothermophilus* NRS 1536/3c and the importance of charged groups for morphogenesis and function. J Bacteriol 1987; 169:2804-9.

21. Sleytr UB, Messner P. Crystalline surface layers in procaryotes. J Bacteriol 1988; 170:2891-97.

22. Koval SF. Paracrystalline protein surface arrays on bacteria, Can J Microbiol 1988; 34:407-14.

23. Beveridge TJ, Graham LL. Surface layers of bacteria. Microbiol Rev 1991; 55:684-705.

24. Messner P, Sleytr UB. Crystalline bacterial cell-surface layers. In: Rose AH, ed. Advances in Microbial Physiology. Vol. 33. London: Academic Press, 1992:213-75.

25. Sára M, Sleytr UB. Relevance of charged groups for the integrity of the S-layer from *Bacillus coagulans* E38-66 and for molecular interactions. J Bacteriol 1993; 175:2248-54.

26. Sleytr UB, Messner P, Pum D et al. Crystalline bacterial cell surface layers. Mol Microbiol 1993; 10:911-16.

27. Beveridge TJ. Bacterial S-layers. Curr Opin Struct Biol 1994; 4:204-12.

28. Jaenicke R, Welsch R, Sára M et al. Stability and self-assembly of the S-layer protein of the cell wall of *Bacillus stearothermophilus*. Hoppe-Seyler's. Z Physiol Chem 1985; 366:663-70.

29. Sleytr UB. Heterologous reattachement of regular arrays of glyco-proteins on bacterial surfaces. Nature 1975; 257:400-2.

30. Sleytr UB, Glauert AM. Analysis of regular arrays of subunits on bacterial surfaces; evidence for a dynamic process of assembly. J Ultrastruct Res 1975; 50:103-16.

31. Sleytr UB, Messner P. Self assembly of crystalline bacterial cell surface layers (S-layers). In: Plattner H, ed. Electron microscopy of subcellular dynamics. Boca Raton: CRC Press, 1989:13-31.

32. Pum D, Messner P, Sleytr UB. The role of the S-layer in the morphogenesis and cell division of the archaebacterium *Methanocorpusculum sinense*. J Bacteriol 1991; 173:6865-73.

33. Sleytr UB, Plohberger R. The dynamic process of assembly of two-dimensional arrays of macromolecules on bacterial cell walls. In: Baumeister W, Vogell W, eds. Electron Microscopy at Molecular Dimensions. Berlin: Springer Verlag, 1980: 36-47.

34. Sleytr UB. Morphopoietic and functional aspects of regular protein membranes present on prokaryotic cell walls. In: Kiermayer O, ed. Cytomorphogenesis in Plants. Cell Biology Monographs. Vol 8. Wien: Springer Verlag, 1981:1-26.

35. Sleytr UB. Self-assembly of the hexagonally and tetragonally arranged subunits of bacterial surface layers and their reattachement to cell walls. J Ultrastruct Res 1976; 55:360-77.

36. Sleytr UB, Sára M, Küpcü Z et al. Structural and chemical characterization of S-layers of selected strains of *Bacillus stearothermophilus* and *Desulfotomaculum nigrificans*. Arch Microbiol 1986; 146:19-24.

37. Sára M, Sleytr UB. Charge distribution on the S-layer of *Bacillus stearothermophilus* NRS 1536/3c and the importance of charged groups for morphogenesis and function. J Bacteriol 1987; 169:2804-9.

38. Sára M, Sleytr UB. Molecular sieving through S-layers of *Bacillus stearothermophilus* strains. J Bacteriol 1987; 169:4092-98.

39. Sára M, Pum D, Sleytr UB. Permeability and charge-dependent adsorption properties of the S-layer lattice from *Bacillus coagulans* E38-66. J Bacteriol 1992; 174:3487-93.

40. Sára M, Küpcü S, Sleytr UB. Localization of the carbohydrate residue of the S-layer glycoprotein from *Clostridium thermohydrosulfuricum* L111-69. Arch Microbiol 1989; 151:416-20.

41. Sára M, Pum D, Küpcü S et al. Isolation of two physiologically induced variant strains of *Bacillus stearothermophilus* NRS 2004/3a and characterization of their S-layer lattices. J Bacteriol 1994; 176:848-60.

42. Sára M, Sleytr UB. Comparative studies of S-layer proteins from *Bacillus stearothermophilus* strains expressed during growth in continuous culture under oxygen-limited and non-oxygen-limited conditions. J Bacteriol 1994; 176:7182-89.

43. Baumeister W, Engelhardt H. Three-dimensional structure of bacterial surface layers. In: Harris JR, Horne RW, eds. Electron microscopy of proteins. Vol. 6. London: Academic Press, 1987:109-54.

44. Baumeister W, Wildhaber I, Engelhardt H. Bacterial surface proteins: some structural, functional and evolutionary aspects. Biophys Chem 1988; 29:39-49.

45. Hovmöller S, Sjögren A, Wang DN. The structure of crystalline bacterial surface layers. Prog Biophys Molec Biol 1988; 51:131-61.

46. Sander LM. Fractal growth processes. Nature 1986; 322:789-93.

47. Ben-Jacob E, Garik P. The formation of patterns in non-equilibrium growth. Nature 1990; 343:523-30.

48. Douglas K, Clark NA. Nanometer molecular lithography. Appl Phys Lett 1986; 48:676-78.

49. Douglas K, Clark NA. Nanometer molecular lithography. In: Carter FL, Siatkowski RE, Wohltjen H, eds. Molecular electronic devices. North-Holland: Elsevier Science Publisher BV, 1988:29-38.

50. Douglas K, Devaud G, Clark NA. Transfer of biologically derived nanometer-scale patterns to smooth substrates. Sience 1992; 257:642-44.

51. Shedd GM, Russell PE. The scanning tunneling microscope as a tool for nanofabrication. Nanotechnol 1990; 1:67-80.

52. Wendel M, Kühn S, Lorenz H et al. Nanolithography with an atomic force microscope for integrated fabrication of quantum electronic devices. Appl Phys Lett 1994; 65:1775-77.

53. Wise KD, Najafi K. Microfabrication techniques for integrated sensors and microsystems. Science 1991; 254:1335-42.

54. Roberts G, ed. Langmuir-Blodgett films. New York: Plenum Press, 1990.

55. Ulman A, ed. Ultrathin organic films: from Langmuir-Blodgett to self-assembly. Boston: Academic Press, 1991.

56. Diederich A, Hödl C, Pum D et al. Reciprocal influence between the protein and lipid components of a lipid-protein membrane model. Colloids and surfaces B: Biointerfaces (submitted).

57. Hönig D, Möbius D. Direct visualisation of monolayers at the air/water interface by Brewster angle microscopy. J Phys Chem 1991; 95:4590-92.

58. Kubalek EW, Kornberg RD, Darst SA. Improved transfer of two-dimensional crystals from the air/water interface to specimen support grids for high-resolution analysis by electron microscopy. Ultramicroscopy 1991; 35:295-304.

59. Asturias FJ, Kornberg RD. A novel method for transfer of two-dimensional crystals from the air water interface to specimen grids— EM sample preparation lipid-layer crystallization. J Struct Biol 1995; 114:60-66.

60. Sára M, Sleytr UB. Use of crystalline bacterial cell envelope layers as ultrafiltration membranes and supports for the immobilization of macromolecules. In: Dechema biotechnology conferences. Vol. 2. Weinheim: VCH Verlagsgesellschaft, 1988:35-51.

61. Sára M, Sleytr UB. Use of regularly structured bacterial cell envelope layers as matrix for the immobilization of macromolecules. Appl Microbiol Biotechnol 1989; 30:184-189.

62. Sára M, Küpcü S, Weiner C et al. Crystalline protein layers as isoporous molecular sieves and immobilization and affinity matrices. In: Sleytr UB, Messner P, Pum D, Sára M, eds. Immobilised macromolecules: application potential. London: Springer Verlag, 1992:71-86.

63. Messner P, Pum D, Sára M et al. Ultrastructure of the cell envelope of the archaebacteria *Thermoproteus tenax* and *Thermoproteus neutrophilus*. J Bacteriol 1986; 166:1046-54.

64. Neubauer A, Pum D, Sleytr UB. An amperometric glucose sensor based on isoporous crystalline protein membranes as immobilization matrix. Anal Lett 1993; 26:1347-60.

65. Neubauer A, Hödl C, Pum D et al. A multistep enzyme sensor for sucrose based on S-layer microparticles as immobilization matrix. Anal Lett 1994; 27:849-65.

66. Neubauer A, Pum D, Sleytr UB et al. Fiber-optic glucose biosensor using enzyme membranes with 2-D crystalline structure. Biosens Bioelectron 1995; (in press)

67. Taga, K, Kellner R, Kainz U et al. In situ attenuated total reflectance FT-IR analysis of an enzyme-modified mid-infrared fiber surface using crystalline bacterial surface proteins. Anal Chem 1993; 66:35-39.

68. Butt HJ, Downing KH, Hansma PK. Imaging the membrane protein bacteriorhodopsin with the atomic force microscope. Biophys J 1990; 58:1473-80.

69. Egger M, Ohnesorge F, Weisenhorn AL et al. Wet lipid-protein membranes imaged at submolecular resolution by atomic force microscopy. J Struct Biol 1990; 103:89-94.

70. Wiegräbe W, Nonnenmacher M, Guckenberger R et al. Atomic force microscopy of a hydrated bacterial surface protein. J Microsc 1991; 163:79-84.

71. Ohnesorge F, Heckl WM, Häberle W et al. Scanning force microscopy studies of the S-layers from *Bacillus coagulans* E38-66, *Bacillus sphaericus* CCM 2177 and of an antibody binding process. Ultramicroscopy 1992; 42-44:1236-42.

72. Devaud G, Furcinitti PS, Fleming JC et al. Direct observation of defect structures in protein crystals by atomic force and transmission electron microscopy. Biophys J 1992; 63:630-38.

73. Firtel M, Southam G, Beveridge TJ et al. Investigation of lattice surface layers by scanning probe microscopy. In: Beveridge TJ, Koval SF, eds. Advances in bacterial paracrystalline surface layers. New York: Plenum Press, 1992; 243-56.

74. Southam G, Firtel M, Blackford BL et al. Transmission electron microscopy, scanning tunneling microscopy, and atomic force microscopy of the cell envelope layer of the archaeobacterium *Methanospirillum hungatei* GP1. J Bacteriol 1993; 175:1946-55.

75. Karrasch S, Hegerl R, Hoh JH et al. Atomic force microscopy produces faithful high resolution images of protein surfaces in an aqueous environment. Proc Natl Acad Sci USA 1994; 91:836-38.

76. Schabert FA, Engel A. Reproducible acquisition of Escherichia coli porin surface topographs by atomic force microscopy. Biophys J 1994; 67:2394-2403.

77. Eigler DM, Schweizer EK. Positioning single atoms with a scanning tunneling microscope Nature 1990; 344:524-26.

78. Eigler DM, Lutz CP, Rudge WE. An atomic switch realized with the scanning tunneling microscope. Nature 1991; 352:600-3.

79. Avouris P. Manipulation of matter at the atomic and molecular levels. Account Chem Res 1995; 28:95-102.

80. Adamchuk VK, Ermakov AV. Device for direct writing and reading-out of information based on the scanning tunneling microscope. Ultramicroscopy 1992; 45:1-4.

81. Sato A, Tsukamoto Y. Nanometre-scale recording and erasing with the scanning tunneling microscope. Nature 1993; 363:431-32.

82. Dulcey CS, Georger JH, Krauthamer V et al. Deep UV photochemistry of chemisorbed monolayers: patterned coplanar molecular assemblies. Science 1991; 252:551-54.

83. Calvert JM. Lithographic patterning of self-assembled films. J Vac Sci Technol B 1993; 11:2155-63.

84. Bhatia SK, Teixeira JL, Anderson M et al. Fabrication of surfaces resistant to protein adsorption and application of two-dimensional protein patterning. Anal Biochem 1993; 208:197-205.

85. Taylor GN, Hutton RS, Stein SM et al. Self-assembly: its use in at-the-surface imaging schemes for microstructure fabrication in resist films. Microelectr Eng 1994; 23:259-62.

86. Vargo TG, Gardella JA, Calvert JM et al. Adhesive electroless metallization of fluoropolymeric substrates. Science 1993; 262:1711-1712.

87. Sleytr UB, Glauert AM. Bacterial cell walls and membranes. In: Harris JR, ed. Electron Microscopy of Proteins. Vol. 3. London: Academic Press, 1982:41-76.

88. Sleytr UB, Messner P. Crystalline surface layers on bacteria. Ann Rev Microbiol 1983; 37:311-39.

89. König H. Archaeobacterial cell envelopes. Can J Microbiol 1988; 34:395-406.

90. Bauer S, Heckmann K, Six L et al. Hyperfiltration through cross-linked monolayers. Desal 1983; 46:369-78.

91. Heckmann K, Strobl Ch, Bauer S. Hyperfiltration through cross-linked monolayers. Thin Solid Films 1983; 99:265-69.

92. Sleytr UB, Sára M. Ultrafiltration membranes with uniform pores from crystalline bacterial cell envelope layers. Appl Microbiol Biotechnol 1986; 25:83-90.

93. Sleytr UB, Sára M. 1985. Structures with membranes having continous pores. United States Patent Nr. 4,752,395.

94. Sleytr UB, Sára M. 1988. Use of structures with membranes having continous pores. United States Patent Nr. 4,849,109.

95. Küpcü S, Sára M, Sleytr UB. Liposomes coated with crystalline bacterial cell surface protein (S-layer) as immobilization structures for macromolecules. Biochim Biophys Acta 1995; 1235:263-69.

96. Racker E. A new procedure for the reconstitution of biologically active phospholipid vesicles. Biochem Biophys Res Comm 1973; 55:224-30.

97. Dencher NA. Spontaneous transmembrane insertion of membrane proteins into lipid vesicles facilitated by short-chain lecithins. Biochem 1986; 25:1195-1200.

98 Mann S. Biomineralization and biomimetic materials chemistry. J Mater Chem 1995; 5:935-46.

99. Mann S. Biominerals and biomimetics: smart solutions to living in the material world. Chemistry & Industry 6 Feb 1995; 93-96.

100. Schulze-Lam S, Harauz G, Beveridge TJ. Participation of a cyanobacterial S-layer in fine grain mineral formation. J Bacteriol 1992; 174:7971-81.

101. Singer RR, Leitner A, Aussenegg F. Structure analysis and models for optical constants of discontinous metallic silver films. J Opt Soc Am B 1995; 12:220-28.

Crystalline Surface Layers on Eubacteria and Archaeobacteria[a]

A survey of S-layer-carrying prokaryotes identified up to 1991, including characterization data and references, has been given in the review by Messner and Sleytr.[1] The following compilation provides an updated list of organisms with citations of the original work.

Organism	Characterization (lattice[b], spacing[c], M_r[d])	Reference
Section 1. Spirochaetes		
Spirochaeta plicatilis		Ref. 1
Spirochaeta stenostrepta Z1		Ref. 1
Spirochaeta zuelzera		Ref. 1
Spirochaeta litoralis R1		Ref. 1
Spirochaeta aurantia J1		Ref. 1
Treponema pallidum Nichols		Ref. 1
Treponema phagedenis, biotype Reiter		Ref. 1
Treponema refringens		Ref. 1
Treponema minutum (CIP 5162)		Ref. 1
Treponema calligyrum (CIP 6441)		Ref. 1
Treponema genitalis VDRL-2		Ref. 1
Treponema microdentium (several strains)		Ref. 1
Treponema sp. E-21		Ref. 1
Section 2. Aerobic/microaerophilic, motile, helical/vibroid Gram-negative bacteria		
Aquaspirillum sinuosum		Ref. 1
Aquaspirillum serpens (several strains)		Ref. 1
Aquaspirillum putridiconchylium (ATCC 15279)		Ref. 1
Aquaspirillum metamorphum (ATCC 15280)		Ref. 1
Aquaspirillum "Ordal"		Ref. 1
Aquaspirillum sp.		Ref. 1
Campylobacter fetus (several strains)		Ref. 1

Organism	Characterization (lattice[b], spacing[c], M_r[d])	Reference
Campylobacter fetus subsp. *fetus* Cf 92-1	-; -; 99	2
Campylobacter pylori		Ref. 1
Campylobacter rectus	-; -; 150	3,4

Section 4. Gram-negative aerobic rods and cocci
Pseudomonadaceae

Pseudomonas putida		Ref. 1
Pseudomonas acidovorans (several strains)		Ref. 1
Pseudomonas delafieldii		Ref. 1
Pseudomonas facilis		Ref. 1
Pseudomonas avenae (NCPPB 1011)		Ref. 1
Pseudomonas-like bacterium strain EU2		Ref. 1

Azotobacteriaceae

Azotobacter vinelandii		Ref. 1
Azomonas agilis		Ref. 1
Azomonas insigne		Ref. 1

Methylococcaceae

Methylomonas albus (several strains)		Ref. 1

Neisseriaceae

Acinetobacter sp. (several strains)		Ref. 1

Other genera

Thermus aquaticus		Ref. 1
Thermus thermophilus HB8 (ATCC 27634)		Ref. 1
Thermomicrobium roseum (ATCC 27502)		Ref. 1
Bordetella pertussis Tohama III		Ref. 1
Lampropedia hyalina		Ref. 1
"*Spinomonas maritima*" D71 (NCMB 2018)	P; -; -	5

Section 5. Facultatively anaerobic rods
Vibrionaceae

Aeromonas salmonicida (several strains)		Ref. 1
Aeromonas salmonicida (several strains)	-; -; 50	6
Aeromonas hydrophila		Ref. 1
Aeromonas hydrophila (several strains)	-; -; ~52	7
Aeromonas hydrophila, serotype O:11 (several strains)	-; -; ~52	8
Aeromonas sobria		Ref. 1
Aeromonas sobria, serotype O:11 (several strains)	-; -; ~52	8
Aeromonas sp. (several strains)	-; -; ~52	9

Other genera

Cardiobacterium hominis		Ref. 1

Section 6. Anaerobic Gram-negative straight, curved and helical rods
Bacteroidaceae

Bacteroides buccae (several strains)		Ref. 1
Bacteroides nodosus (several strains)		Ref. 1

Organism	Characterization (lattice[b], spacing[c], M_r[d])	Reference
Bacteroides capillus (ATCC 33690, ATCC 33691)		Ref. 1
Bacteroides pentasaceus NP333 and WPH61		Ref. 1
Bacteroides forsythus (several strains)		Ref. 1
Bacteroides heparinolyticus		Ref. 1
Bacteroides sp. (several strains)		Ref. 1
Porphyromonas spp. (formerly *Bacteroides* ssp.)	P; –; –	10
Wolinella recta (several strains)		Ref. 1
Selenomonas palpitans		Ref. 1

Section 9. Rickettsiae and chlamydiae

Rickettsiaceae

Organism	Characterization	Reference
Rickettsia prowazekii (several strains)		Ref. 1
Rickettsia typhi		Ref. 1
Rickettsia rickettsii		Ref. 1
Rickettsia akari		Ref. 1

Chlamydiaceae

Organism	Characterization	Reference
Chlamydia trachomatis TE55		Ref. 1
Chlamydia psittaci		Ref. 1

Section 12. Gram-positive cocci

Organism	Characterization	Reference
Deinococcus radiodurans (several strains)		Ref. 1
Peptostreptococcus asaccharolyticus (ATCC 14963)		Ref. 1
Peptostreptococcus magnus AHC 5155		Ref. 1
Peptostreptococcus anaerobius (ATCC 27337)	P; –; 78	11

Section 13. Endospore-forming Gram-positive rods and cocci

Organism	Characterization	Reference
Bacillus subtilis		Ref. 1
Paenibacillus alvei (formerly *Bacillus alvei*) (several strains)		Ref. 1
Bacillus anthracis		Ref. 1
Bacillus anthracis 9131	p6; –; 87.3	12
Bacillus anthracis Delta Sterne-1	P; 8-10; 95	13
Bacillus brevis (several strains)		Ref. 1
Bacillus brevis (several strains)	–; –; 110-150	14
Bacillus cereus (several strains)		Ref. 1
Bacillus cereus GA 682	–; –; –	15
Bacillus circulans (several strains)		Ref. 1
Bacillus coagulans E38-66		Ref. 1
Bacillus fastidiosus		Ref. 1
Bacillus licheniformis NM 105		Ref. 1
Bacillus megaterium		Ref. 1
Bacillus polymyxa (several strains)		Ref. 1
Bacillus schlegelii (DSM 2000)		Ref. 1
Bacillus sphaericus (several strains)		Ref. 1
Bacillus stearothermophilus (several strains)		Ref. 1
Bacillus thuringiensis ssp. *galleriae* (NRLL 4045)		Ref. 1
Bacillus aneurinolyticus (several strains)		Ref. 1

Organism	Characterization (lattice[b], spacing[c], M$_r$[d])	Reference
Bacillus aneurinolyticus (several strains)	-; -; 105, 115	16
Bacillus macroides A and D		Ref. 1
Bacillus psychrophilus W16A		Ref. 1
Bacillus sp. (several strains)		Ref. 1
Bacillus borstelensis (2 strains)	-; -; 125-150	17
Bacillus choshinensis (2 strains)	-; -; 135	14
Bacillus formosus (2 strains)	-; -; ~150	17
Bacillus galactophilus (2 strains)	-; -; ~150	14
Bacillus migulanus HP926	-; -; ~150	14
Bacillus parabrevis (several strains)	-; -; 110-150	14
Bacillus reuszeri (3 strains)	-; -; 120-180	17
Bacillus thermoaerophilus L420-91 DSM 10154)	p4; 10; 116	18
Sulfobacillus thermosulfidooxidans	P; -; -	19
Clostridium aceticum (several strains)		Ref. 1
Clostridium aminovalericum T2-7		Ref. 1
Clostridium viride (formerly *C. aminovalericum*)	p6; 18.5; 110	20
Clostridium botulinum		Ref. 1
Clostridium botulinum type E Saroma	-; -; 10-150	21
Clostridium difficile (several strains)		Ref. 1
Clostridium difficile (GAI 1152)	p4; 7.8; 38, 42	22
Clostridium formicoaceticum (DSM 912)		Ref. 1
Clostridium novyi		Ref. 1
Clostridium polysaccharolyticum		Ref. 1
Clostridium sporogenes		Ref. 1
Clostridium symbiosum HB25		Ref. 1
Clostridium tetani		Ref. 1
Clostridium tetani AO 174	-; -; 140	23
Clostridium thermoautotrophicum		Ref. 1
Clostridium thermocellum	-; -; -	24
Thermoanaerobacter thermohydrosulfuricus (several strains) (formerly *Clostridium thermohydrosulfuricum*)		Ref. 1
Thermoanaerobacter kivui (formerly *Acetogenium kivui*)		Ref. 1
Clostridium thermosaccharolyticum (several strains)		Ref. 1
Clostridium tyrobutyricum (several strains)		Ref. 1
Clostridium lentoputrescens (ATCC 17791)		Ref. 1
Clostridium tartarivorum (several strains)		Ref. 1
Clostridium thermolacticum (DSM 2911)		Ref. 1
Clostridium xylanolyticum (ATCC 49623)		Ref. 1
Clostridium xylanolyticum (ATCC 49623)	p2; 6.6/5.3; 180	25
Clostridium uzonii 1501/60	p6; -; -	26
Clostridium sp. EM1		Ref. 1
St. Lucia strain SLH	p6; 21; 200	27
Desulfotomaculum nigrificans (several strains)		Ref. 1
Desulfotomaculum australicum AB33 (ACM 3917)	P; -; -	28
St. Lucia strain SLT	p4; 11; 83	27
Sporosarcina ureae (ATCC 13881)		Ref. 1

Organism	Characterization (lattice[b], spacing[c], M$_r$[d])	Reference
Section 14. Regular, non-sporing, Gram-positive rods		
Lactobacillus acidophilus (several strains)		Ref. 1
Lactobacillus acidophilus (JCM 1132)	–; –; 43	29
Lactobacillus crispatus (JCM 5810)	–; –; 43	29
Lactobacillus helveticus (several strains)		Ref. 1
Lactobacillus helveticus (ATCC 12046)	p2; 9.6/4.5; 52	30
Lactobacillus casei		Ref. 1
Lactobacillus plantarum 41021/252	p2; 8.3/5.3; 56	31
Lactobacillus brevis (several strains)		Ref. 1
Lactobacillus buchneri (several strains)		Ref. 1
Lactobacillus buchneri 41021/251	p2; 6.1/5.4; 53	31
Lactobacillus fermentum (NCTC 7230)		Ref. 1
Lactobacillus bulgaricus (YIT 0045)		Ref. 1
Lactobacillus sp.		Ref. 1
Enteric lactobacilli (several strains)		Ref. 1
Section 15. Irregular, non-sporing, Gram-positive rods		
Corynebacterium diphteriae C4		Ref. 1
Corynebacterium glutamicum	p6; –; 55	32
Group JK bacteria (coryneform rods)		Ref. 1
Propionibacterium freudenreichii (CNRZ 722)	p2; 9.9/5.4; 57	33
Propionibacterium jensenii (CNRZ 87)	p2; 9.9/5.4; 67	33
Eubacterium tenue		Ref. 1
Eubacterium yurii (several strains)		Ref. 1
Eubacterium lentum AHP 6099		Ref. 1
Eubacterium sp. AHN 990		Ref. 1
Acetobacterium woodii WB1		Ref. 1
Thermoanaerobacter ethanolicus		Ref. 1
Section 16. Mycobacteria		
Mycobacterium bovis BCG (five strains)		Ref. 1
Section 18. Anoxigenic phototrophic bacteria		
Purple bacteria		
Chromatiaceae		
Chromatium okenii		Ref. 1
Chromatium weissei		Ref. 1
Chromatium warmingii		Ref. 1
Chromatium buderi		Ref. 1
Chromatium gracile		Ref. 1
Thiocapsa floridana 9314		Ref. 1
Ameobobacter bacillosus		Ref. 1
Ectothiorhodospiraceae		
Ectothiorhodospira mobilis 8112		Ref. 1
Ectothiorhodospira halochloris BN9850		Ref. 1

Organism	Characterization (lattice[b], spacing[c], M_r[d])	Reference
Rhodospirillaceae		
Rhodospirillum rubrum		Ref. 1
Rhodospirillum molischianum		Ref. 1
Rhodospirillum salexigens (DSM 2132)		Ref. 1
Rhodopseudomonas palustris		Ref. 1
Rhodopseudomonas acidophila		Ref. 1
Green bacteria		
Pelodictyon sp.		Ref. 1
Chlorochromatium aggregatum		Ref. 1

Section 19. Oxygenic photosynthetic bacteria

Cyanobacteria		
Chroococcales		
Synechococcus sp. GC		Ref. 1
Cyanothece minerva		Ref. 1
Aphanothece halophytica (ATCC 29534)		Ref. 1
Gloeocapsa alpicola		Ref. 1
Aphanocapsa rivularis		Ref. 1
Aphanocapsa sp.		Ref. 1
Merismopedia sp.		Ref. 1
Synechocystis aquatilis (several strains)		Ref. 1
Synechocystis fuscopigmentosa		Ref. 1
Synechocystis sp. PCC 6803		Ref. 1
Synechocystis sp. (several strains)		Ref. 1
Microcystis firma		Ref. 1
Microcystis firma (several strains)	p6; 14.3-16.1; –	34
Microcystis incerta		Ref. 1
Microcystis marginata		Ref. 1
Microcystis sp.		Ref. 1
Chroococcacean cyanobacteria (22 strains)		Ref. 1
Pleurocapsales		
Chroococcidiopsis sp.		Ref. 1
Oscillatoria princeps	p4; –; –	35
Phormidium uncinatum (2 strains)	p4; –; –	35
Lyngbya aeruginosa	p4; –; –	35

Section 20. Chemolithotrophic bacteria and associated organisms

Nitrifying bacteria		
Nitrosomonas sp.		Ref. 1
Nitrosospira sp. X101		Ref. 1
Nitrosocystis oceanus		Ref. 1
Colourless sulphur bacteria		
Thiobacillus kabobis		Ref. 1

Organism	*Characterization* (lattice[b], spacing[c], M_r^d)	*Reference*

Section 21. Budding and/or appendaged bacteria

Prosthecate bacteria

Hyphomicrobium-type halophilic microorganism		Ref. 1
Pedomicrobium sp.		Ref. 1
Caulobacter crescentus CB15		Ref. 1
Caulobacter sp. (several strains)		Ref. 1
Freshwater *Caulobacter* sp. (several strains)	p6; –; 100-193	36

Non-prosthecate bacteria

Planctomyces sp. ES		Ref. 1
"*Planctomyces gracilis*" Hortobágyi 1965		Ref. 1

Section 23. Non-photosynthetic, non-fruiting gliding bacteria

Cytophagaceae

Cytophaga johnsonae (ATCC 17061)		Ref. 1

Other genera

Flexibacter columnaris		Ref. 1
Flexibacter polymorphus (ATCC 27820)		Ref. 1

Section 24. Myxobacteria

Myxococcus xanthus DK 1050		Ref. 1

Section 25. Archaeobacteria[a]

Crenarchaeota

Thermoproteales

Thermoproteus tenax Kra 1 (DSM 2078)		Ref. 1
Thermoproteus neutrophilus Hvv24 (DSM 2338)		Ref. 1
Thermoproteus autotrophicus		Ref. 1
Thermoproteus uzoniensis Z-605 (DSM 5263)	p6; –; –	37
Thermoproteus sp. H3		Ref. 1
Pyrobaculum organotrophum H10		Ref. 1
Pyrobaculum organotrophum inner layer	p6; 27.9; –	38,39
(DSM 4185) outer layer	p6; 20.6; 80	
Pyrobaculum islandicum GEO3 (DSM 4184)		Ref. 1
Pyrobaculum aerophilum IM2 (DSM 7523)	p6; 30; –	40
Thermofilum pendens Hvv3 (DSM 2475)		Ref. 1
Desulfurococcus mucosus (DSM 2162)		Ref. 1
Desulfurococcus mobilis (DSM 2161)		Ref. 1
Staphylothermus marinus F1 (DSM 3639)		Ref. 1
Staphylothermus marinus F1 (DSM 3639)	p6; –; 10-130	41

"Pyrodictiales"

Pyrodictium occultum PL-19 (DSM 2709)		Ref. 1
Pyrodictium brockii s1 (DSM 2708)		Ref. 1
Pyrodictium abyssi AV2 (DSM 6158)	p6; 15; 126	42
Hyperthermus butylicus		Ref. 1
Thermodiscus maritimus S2		Ref. 1

Organism	Characterization (lattice[b], spacing[c], M$_r$[d])	Reference
Sulfolobales		
Sulfolobus acidocaldarius (several strains)		Ref. 1
Sulfolobus solfataricus-"*Caldariella acidophila*" (DSM 1616, DSM 1617)		Ref. 1
Sulfolobus shibatae B12 (DSM 5389)	p3; 20.5; –	43,44
Sulfolobus metallicus Kra 23 (DSM 6482)	P; –; –	45
Acidianus infernus So4a (DSM 3191)		Ref. 1
Acidianus brierleyi (DSM 1651)		Ref. 1
Acidianus brierleyi (formerly *Sulfolobus brierleyi*)	p3; 19; –	46
Metallosphaera sedula TH2 (DSM 5348)	p6; –; –	47
Stygiolobus azoricus FC6 (DSM 6296)		Ref. 1
Sulfurococcus mirabilis AT-59	p6; –; –	48,49
Sulfurococcus yellowstonii Str6kar	p6; –; –	49,50
Euryarchaeota		
Thermococcales		
Thermococcus celer (DSM 2476)		Ref. 1
Thermococcus stetteri K-15 (DSM 5262)	p6; 18; 210, 80	51,52
Thermococcus litoralis NS-C (DSM 5473)	P; –; –	53
Pyrococcus furiosus Vc1 (DSM 3638)		Ref. 1
Pyrococcus abyssi GE5 (CNCM 1302)	P; –; –	54
"Archaeoglobales"		
Archaeoglobus fulgidus (several strains)		Ref. 1
Archaeoglobus profundus AV 18 (DSM 5631)		Ref. 1
Methanococcales		
Methanococcus vannielii SB (DSM 1224)		Ref. 1
Methanococcus voltae PS (DSM 1537)		Ref. 1
Methanococcus thermolithotrophicus SN1 (DSM 2095)		Ref. 1
Methanococcus jannaschii JAL-1 (DSM 2661)		Ref. 1
Methanococcus aeolicus PL-15/H		Ref. 1
Methanococcus igneus Kol 5 (DSM 5666)		Ref. 1
Methanobacteriales		
Methanobacterium sp. G2R		Ref. 1
Methanothermus fervidus V24S (DSM 2088)		Ref. 1
Methanothermus sociabilis KF1-F1		Ref. 1
Methanomicrobiales		
Methanomicrobium mobile BP		Ref. 1
Methanococcoides methylutens TMA-10 (ATCC 33938)		Ref. 1
Methanogenium cariaci JR1 (DSM 1497)		Ref. 1
Methanogenium marisnigri JR1 (DSM 1498)		Ref. 1
Methanoculleus marisnigri JR1	p6; 13.5; 138	55
Methanogenium thermophilicum (DSM 2640)		Ref. 1
Methanogenium tationis (DSM 2702)		Ref. 1
Methanogenium liminatans (DSM 4140)		Ref. 1
Methanoculleus bourgense (formerly *Methanogenium bourgense*)		56,57
Methanocorpusculum parvum XII (DSM 3823)		Ref. 1
Methanocorpusculum sinense (DSM 4274)		Ref. 1
Methanocorpusculum bavaricum (DSM 4179)		Ref. 1

Organism	Characterization (lattice[b], spacing[c], M_r[d])	Reference
Methanocorpusculum labreanum Z (DSM 4855)	p6; ~25; –	58
Methanohalophilus oregonense WAL1 (DSM 5435)		Ref. 1
Methanolobus tindarius T3 (DSM 2278)		Ref. 1
Methanolobus siciliae T4/M (DSM 3028)		Ref. 1
Methanolacinia paynteri (DSM 2545)		Ref. 1
Methanoplanus limicola M3 (DSM 2279)		Ref. 1
Methanoplanus limicola (DSM 2279)	p6; 14.7; 135	59
Methanothrix soehngenii (DSM 2139)		Ref. 1
Methanothrix concilii		Ref. 1
Methanosaeta concilii (formerly *Methanothrix concilii*)		60
Methanosarcina mazei (DSM 2053)		Ref. 1
Methanosarcina acetivorans C2A (DSM 2834, ATCC 35395)		Ref. 1
Methanosarcina sp. (several strains)	P; –; –	61
Methanospirillum hungatei (several strains)		Ref. 1
Methanopyrales		
Methanopyrus kandleri AV19 (DSM 6324)	P; –; –	62
Halobacteriales		
Halobacterium salinarium (DSM 668, ATCC 19700)		Ref. 1
"*Halobacterium halobium*" (several strains)		Ref. 1
"*Halobacterium cutirubrum*"		Ref. 1
Halobacterium saccharovorum (DSM 1137)		Ref. 1
Halobacterium sp. (several strains)		Ref. 1
Haloarcula japonica TR-1	p6; 10; 180	63,64
Haloferax volcanii (several strains)		Ref. 1

Isolates with no taxonomic affiliation

Eubacteria:

Pelobacter carbinolicus (DSM 2909)		Ref. 1
Thermotoga maritima (DSM 3109)		Ref. 1
Desulfurella acetivorans A63	P; –; –	65
Desulfurella multipotens RH-8	p6; –; –	66
Carboxydothermus hydrogenoformans (DSM 6008)	P; –; –	67
Nitriloacetate-utilizing bacteria (several strains)		Ref. 1
Putrescine-degrading bacterium (strain NorPut1)		Ref. 1

Archaeobacteria:

"Square bacterium" (halophilic)		Ref. 1
Sulfosphaerellus thermoacidophilum S-5	p6; 15.3; -110	68,69
Thermophilic archaebacterium, strain ES1		Ref. 1
Chemoorganotrophic archaeobacterium ES4	P; –; –	70

---

[a] Classification according to Bergey's Manual of Systematic Bacteriology, Vols 1-4[71] and Section 25. Archaeobacteria, according to Kandler.[72]

[b] Unit cell space group symmetry; p2, p3, p4, p6; P, periodic structure not further characterized (e.g. evidence from thin sections).

[c] Center-to-center spacing (nm).

[d] Molecular mass of the S-layer (glyco)protein (kDa).

REFERENCES

1. Messner P, Sleytr UB. Crystalline bacterial cell-surface layers. In: Rose AH, ed. Advances in Microbial Physiology. Vol. 33. London: Academic Press, 1992:213-75.

2. Grollier G, Burucoa C, Ricco JB et al. Isolation and immunogenicity of *Campylobacter fetus* subsp. *fetus* from an abdominal aortic aneurysm. Eur J Clin Microbiol Infect Dis 1993; 12:847-49.

3. Kaneko T. Analysis of cell surface antigens of *Campylobacter rectus*. Bull Tokyo Dental College 1992; 33:171-85.

4. Kobayashi Y, Ohta H, Kokeguchi S et al. Antigenic properties of *Campylobacter rectus* (*Wolinella recta*) major S-layer proteins. FEMS Microbiol Lett 1993; 108:275-80.

5. Bayer ME, Easterbrook K. Tubular spinae are long-distance connectors between bacteria. J Gen Microbiol 1991; 137:1081-86.

6. Fernández AIG, Pérez MJ, Rodríguez LA et al. Surface phenotypic characteristics and virulence of Spanish isolates of *Aeromonas salmonicida* after passage through fish. Appl Environ Microbiol 1995; 61:2010-12.

7. Sakata T, Shimojo T. Surface structure and pathogenicity of *Aeromonas hydrophila* strains isolated from diseased and healthy fish. Memoirs of the Faculty of Fisheries Kagoshima University 1991; 40:47-58.

8. Merino S, Camprubi S, Tomas JM. Self-pelleting autoagglutination on mesophilic aeromonads belonging to serotype O:11. World J Microbiol Biotechnol 1991; 7:276-78.

9. Ford LA, Thune RL. S-layer positive motile aeromonads isolated from channel catfish. J Wildlife Dis 1991; 27:557-61.

10. Haapasalo M, Kerosuo E, Lounatmaa K. Hydrophobicities of human polymorphonuclear leukocytes and oral *Bacteroides* and *Porphyromonas* spp., *Wolinella recta*, and *Eubacterium yurii* with special reference to bacterial surface structures. Scand J Dent Res 1990; 98:472-81.

11. Kotiranta A, Haapasalo M, Lounatmaa K et al. Crystalline surface protein of *Peptostreptococcus anaerobius*. Microbiology 1995; 140:1065-73.

12. Etienne-Toumelin I, Sirard J-C, Duflot E et al. Characterization of the *Bacillus anthracis* S-layer: cloning and sequencing of the structural gene. J Bacteriol 1995; 177:614-20.

13. Farchaus JW, Ribot WJ, Downs MB et al. Purification and characterization of the major surface array protein from the avirulent *Bacillus anthracis* Delta Sterne-1. J Bacteriol 1995; 177:2481-89.

14. Takagi H, Shida O, Kadowaki K et al. Characterization of *Bacillus brevis* with descriptions of *Bacillus migulanus* sp. nov., *Bacillus choshinensis* sp. nov., *Bacillus parabrevis* sp. nov., and *Bacillus galactophilus* sp. nov. Int J Syst Bacteriol 1993; 43:221-31.

15. Goderdzishvili MG, Gosteva VV, Klitsunova NV et al. Changes in the ultrastructure of *Bacillus cereus* cell wall, determined by plasmid RP4Mucts62. Zh Mikrobiol Epidemiol Immunbiol 1989; 2:8-11.

16. Shida O, Takagi H, Kadowaki K et al. *Bacillus aneurinolyticus* sp. nov., nom. rev. Int J Syst Bacteriol 1994; 44:143-50.

17. Shida O, Takagi H, Kadowaki K et al. Proposal of *Bacillus reuszeri* sp. nov., *Bacillus formosus* sp. nov., nom. rev., and *Bacillus borstelensis* sp. nov., nom. rev. Int J Syst Bacteriol 1995; 45:93-100.

18. Meier-Stauffer K, Busse H-J, Rainey FA et al. Description of the sugar beet isolates *Bacillus thermoaerophilus* sp. nov., emended description of *Bacillus brevis* ATCC 12990 and transfer of this strain to the newly described species. Int J Syst Bacteriol; submitted.

19. Serevina LO, Senyushkin AA, Karavaiko GA. Ultrastructure and chemical composition of the S-layer of *Sulfobacillus thermosulfidooxidans*. Dokl Akad Nauk 1993; 328:633-36.

20. Buckel W, Janssen PH, Schuhmann A et al. *Clostridium viride* sp. nov., a strictly anaerobic bacterium using 5-aminovalerate as growth substrate, previously assigned to *Clostridium aminovalericum*. Arch Microbiol 1994; 162:387-94.

21. Takumi K, Ichiyanagi S, Endo Y et al. Characterization, self-assembly and reattachment of S-layer from *Clostridium botulinum* type E Saroma.Tokushima J Exp Med 1992; 39:101-7.

22. Hagiya H, Oka T, Tsuji H et al. The S-layer composed of two different protein subunits from *Clostridium difficile* GAI 1152: a simple purification method and characterization. J Gen Appl Microbiol 1992; 38:63-74.

23. Takumi K, Susami Y, Takeoka A et al. S-layer protein of *Clostridium tetani*: purification and properties. Microbiol Immunol 1991; 35:569-76.

24. Fujino T, Beguin P, Aubert JP. Organization of a *Clostridium thermocellum* gene cluster encoding the cellulosomal scaffolding protein CipA and a protein possibly involved in attachment of the cellulosome to the cell surface. J Bacteriol 1993; 175:1891-99.

25. Rogers GM, Messner P. Improved description of the cell wall architecture of the xylanolytic eubacterium *Clostridium xylanolyticum*. Int J Syst Bacteriol 1992; 42:492-93.

26. Krivenko VV, Vadachloriya RM, Chermykh NA et al. *Clostridium uzonii*, sp. nov., an anaerobic thermophilic glycolytic bacterium from the hot springs in the Kamchatka peninsula. Mikrobiologiya 1990; 59:1058-66.

27. Karnauchow TM, Koval SF, Jarrell KF. Isolation and characterization of three thermophilic anaerobes from a St. Lucia hot spring. System Appl Microbiol 1992; 15:296-310.

28. Love CA, Patel BKC, Nichols PD et al. *Desulfotomaculum australicum,* sp. nov., a thermophilic sulfate-reducing bacterium from the Great Artesian Basin of Australia. System Appl Microbiol 1993; 16:244-51.

29. Toba T, Virkola R, Westerlund B et al. A collagen-binding S-layer protein in *Lactobacillus crispatus.* Appl Environ Microbiol 1995; 61:2467-71.

30. Lortal S, van Heijenoort J, Gruber K et al. S-layer of *Lactobacillus helveticus* ATCC 12046: isolation, chemical characterization and reformation after extraction with lithium chloride. J Gen Microbiol 1992; 138:611-18.

31. Möschl A, Schäffer C, Sleytr UB et al. Characterization of the S-layer glycoproteins of two lactobacilli. In: Beveridge TJ, Koval SF, eds. Advances in Bacterial Paracrystalline Surface Layers. New York: Plenum, 1993:281-84.

32. Peyret JL, Bayan N, Joliff G et al. Characterization of the *cspB* gene encoding PS2, an ordered surface-layer protein in *Corynebacterium glutamicum.* Mol Microbiol 1993; 9:97-109.

33. Lortal S, Rouault A, Cesselin B et al. Paracrystalline surface layers of dairy propionibacteria. Appl Environ Bacteriol 1993; 59:2369-74.

34. Šmarda J, Komrska J. Advances in S-layer research of chroococcal cyanobacteria. In: Beveridge TJ, Koval SF, eds. Advances in Bacterial Paracrystalline Surface Layers. New York: Plenum, 1993:77-84.

35. Hoiczyk E, Baumeister W. Envelope structure of four gliding filamentous cyanobacteria. J Bacteriol 1995; 177:2387-95.

36. Walker SG, Smith SH, Smit J. Isolation and comparison of the paracrystalline surface layer proteins of freshwater caulobacters. J Bacteriol 1992; 174:1783-92.

37. Bonch-Osmolovskaya EA, Miroshnichenko ML, Kostrikina, NA et al. *Thermoproteus uzoniensis* sp. nov., a new extremely thermophilic archaebacterium from Kamchatka continental hot springs. Arch Microbiol 1990; 154:556-59.

38. Huber H, Kristjansson JK, Stetter KO. *Pyrobaculum* gen. nov., a new genus of neutrophilic, rod-shaped archaebacteria from continental solfataras growing optimally at 100°C. Arch Microbiol 1987; 149:95-101.

39. Phipps BM, Huber R, Baumeister W. The cell envelope of the hyperthermohilic archaebacterium *Pyrobaculum organotrophum* consists of two regularly arrayed protein layers: three-dimensional structure of the outer layer. Mol Microbiol 1991; 5:253-65.

40. Völkl P, Huber R, Drobner E et al. *Pyrobaculum aerophilum* sp. nov., a novel nitrate-reducing hyperthermophilic archaeum. Appl Environ Microbiol 1993; 59:2918-26.

41. Peters J, Nitsch M, Kühlmorgen B et al. Tetrabrachion: a filamentous archae bacterial surface protein assembly of unusual structure and extreme stability. J Mol Biol 1995; 245:385-401.

42. Pley U, Schipka J, Gambacorta A et al. *Pyrodictium abyssi* sp. nov. represents a novel heterotrophic marine archaeal hyperthermophile growing at 110°C. System Appl Microbiol 1991; 14:245-53.

43. Grogan D, Palm P, Zillig W. Isolate B12, which harbours a virus-like element, represents a new species of the archaebacterial genus *Sulfolobus, Sulfolobus shibatae,* sp. nov. Arch Microbiol 1990; 154:594-99.

44. Lembcke G, Baumeister W, Beckmann E et al. Cryo-electron microscopy of the surface protein of *Sulfolobus shibatae.* Ultramicroscopy 1993; 49:397-406.

45. Huber G, Stetter KO. *Sulfolobus metallicus,* sp. nov., a novel strictly chemolithoautotrophic thermophilic archaeal species of metal-mobilizers. System Appl Microbiol 1991; 14:372-78.

46. Baumeister W, Volker S, Santarius U. The three-dimensional structure of the surface protein of *Acidianus brierleyi* determined by electron crystallography. System Appl Microbiol 1991; 14:103-10.

47. Huber G, Spinnler C, Gambacorta A et al. *Metallosphaera sedula* gen. and sp. nov. represents a new genus of aerobic, metal-mobilizing, thermoacidophilic archaebacteria. System Appl Microbiol 1989; 12:38-47.

48. Bashkatova NA, Severina LO, Golovacheva RS et al. Surface layers of extremely thermoacidophilic archaebacteria of the genus *Sulfolobus.* Mikrobiologiya 1991; 60:90-94.

49. Karavaiko GI, Golovacheva RS, Golyshina OV et al. *Sulfurococcus*—a new genus of thermoacidophilic archaebacteria oxidizing sulfur, ferrous iron and sulfide minerals. In: Torma AE, Wey JE, Lakshmanan VL, eds. Biohydrometallurgical Technologies. Warrendale, PA: The Minerals, Metals & Materials Society, 1993:685-94.

50. Karavaiko GI, Golyshina OV, Troitskii AV et al. *Sulfurococcus yellowstonii* sp. nov., a new species of iron- and sulfur-oxidizing thermoacidophilic archaebacteria. Mikrobiologiya 1994; 63:668-82.

51. Miroshnichenko ML, Bonch-Osmolovskaya EA, Neuner A et al. *Thermococcus stetteri* sp. nov., a new extremely thermophilic marine sulfur-metabolizing archaebacterium. System Appl Microbiol 1989; 12:257-62.

52. Gongadze GM, Kostyukova AS, Miroshnichenko ML et al. Regular proteinaceous layers of *Thermococcus stetteri* cell envelope. Curr Microbiol 1993; 27:5-9.

53. Neuner A, Jannasch HW, Belkin S et al. *Thermococcus litoralis* sp. nov.: a new species of extremely thermophilic marine archaebacteria. Arch Microbiol 1990; 153:205-7.

54. Erauso G, Reysenbach A-L, Godfroy A et al. *Pyrococcus abyssi* sp. nov., a new hyperthermophilic archaeon isolated from a deep-sea hydrothermal vent. Arch Microbiol 1993; 160:338-49.

55. Bayley DP, Koval SF. Membrane association and isolation of the S-layer protein of *Methanoculleus marisnigri.* Can J Microbiol 1994; 40:237-41.

56. Maestrojuán GM, Boone DR, Xun L et al. Transfer of *Methanogenium bourgense, Methanogenium marisnigri, Methanogenium olentangyi,* and *Methanogenium thermophilicum* to the genus *Methanoculleus* gen. nov., emendation of *Methanoculleus marisnigri* and *Methanogenium,* and description of new strains of *Methanoculleus bourgense* and *Methanoculleus marisnigri.* Int J Syst Bacteriol 1990; 40:117-22.

57. Ollivier BM, Mah RA, Garcia JL et al. Isolation and characterization of *Methanogenium bourgense* sp. nov. Int J Syst Bacteriol 1986; 36:297-301.

58. Zhao Y, Boone DR, Mah RA et al. Isolation and characterization of *Methanocorpusculum labreanum* sp. nov. from the LaBrea tar pits. Int J Syst Bacteriol 1989; 39:10-13.

59. Cheong G-W, Cejka Z, Peters J et al. The surface protein layer of *Methanoplanus limicola:* three-dimensional structure and chemical characterization. System Appl Microbiol 1991; 14:209-17.

60. Patel GB, Sprott GD. *Methanosaeta concilii* gen. nov., sp. nov. ("*Methanothrix concilii*") and *Methanosaeta thermoacetophila* nom. rev., comb. nov. Int J Syst Bacteriol 1990; 40:79-82.

61. Sower KR, Boone JE, Gunsalus RP. Disaggregation of *Methanosarcina* spp. and growth as single cells at elevated osmolarity. Appl Environ Microbiol 1993; 59:3832-39.

62. Kurr M, Huber R, König H et al. *Methanopyrus kandleri,* gen. and sp. nov. represents a novel group of hyperthermophilic methanogens, growing at 110°C. Arch Microbiol 1991; 156:239-47.

63. Nishiyama Y, Takashina T, Grant WD et al. Ultrastructure of the cell wall of the triangular halophilic archaebacterium *Haloarcula japonica* strain TR-1. FEMS Microbiol Lett 1992; 99:43-48.

64. Nakamura S, Mizutani S, Wakai H et al. Purification and partial characterization of cell surface glycoprotein from extremely halophilic archaeon *Haloarcula japonica* strain TR-1. Biotechnol Lett 1995; 17:705-6.

65. Bonch-Osmolovskaya EA, Sokolova TG, Kostrikina NA et al. *Desulfurella acetivorans* gen. nov. and sp. nov.—a new thermophilic sulfur-reducing eubacterium. Arch Microbiol 1990; 153:151-55.

66. Miroshnichenko ML, Gongadze GA, Lysenko AM et al. *Desulfurella multipotens* sp. nov., a new sulfur-respiring thermophilic eubacterium from Raoul Island (Kermadec archipelago, New Zealand). Arch Microbiol 1994; 161:88-93.

67. Svetlichny VA, Sokolova TG, Gerhardt M et al. *Carboxydothermus hydrogenoformans* gen. nov., sp. nov., a CO-utilizing thermophilic anaerobic bacterium from hydrothermal environments of Kunashir Island. System Appl Microbiol 1991; 14:254-60.

68. Li Y, Xu Y, Cai W et al. Archaebacterial characteristics of *Sulfosphaerellus thermoacidophilum.* Acta Microbiol Sinica 1988; 28:109-14.

69. Xu Y, Li Y, Cai W et al. The biochemical properties of the cell envelope of *Sulfosphaerellus thermoacidophilum.* Acta Microbiol Sinica 1988; 28:221-25.

70. Pledger RJ, Baross JA. Preliminary description and nutritional characterization of a chemoorganotrophic archaeobacterium growing at temperatures of up to 110°C isolated from a submarine hydrothermal vent environment. J Gen Microbiol 1991; 137:203-11.

71. Holt JG, et al, eds. Bergey's Manual of Systematic Bacteriology, Vols. 1-4. Baltimore: Williams & Wilkins, 1989.

72. Kandler O. Archaea (Archaeobacteria). Progr Botany 1993; 54:1-24.

INDEX

Page numbers in italics denote figures (f) or tables (t).

INDEX

Page numbers in italics denote figures (f) or tables (t).

Printed and bound by CPI Group (UK) Ltd, Croydon, CR0 4YY

08/05/2025

01864992-0001